TUTORIAL CHEMISTRY TEXTS

23
Mechanisms in Organic Reactions

RICHARD A. JACKSON
University of Sussex

RS•C
ROYAL SOCIETY OF CHEMISTRY

ISBN 0-85404-642-9

A catalogue record for this book is available from the British Library

Published by The Royal Society of Chemistry, Thomas Graham House, Science Park,
Milton Road, Cambridge CB4 0WF, UK
Registered Charity No. 207890
For further information see our web site at www.rsc.org

Typeset in Great Britain by Alden Bookset, Northampton
Printed and bound by Italy by Rotolito Lombarda

Preface

The wonderful complexity of organic chemistry involves thousands of different reactions which allow the synthesis and interconversions of millions of compounds, some of great complexity. The key to understanding this vital branch of chemistry is the concept of the reaction mechanism.

This book starts with a discussion of how covalent bonds break and form, and how these bond-breaking and bond-forming processes provide the basis of reaction mechanisms. The principles governing how to make sensible suggestions about possible mechanisms are set out, and the distinction is made between elementary reactions, which involve just one step, and stepwise reactions which have more than one step and involve the production of intermediates that react further.

Chapter 2 covers kinetics, which provides useful information about reaction mechanisms, and allows us to distinguish between possible mechanisms in many cases. Elementary reactions do not involve intermediates, but go through a transition state. Although this transition state cannot be isolated, it can be studied in various ways which provide insights into the reaction mechanism, and this forms the subject matter of Chapter 3. This is followed by three chapters on the most important intermediates in organic chemistry: anions, radicals and cations. A final chapter on molecular reactions concerns thermal and photochemical processes. The concepts of frontier orbitals and the aromatic transition state allow us to predict which reactions are "allowed" and which are "forbidden", and provide insights into why most reactions of practical interest involve multi-step processes.

Where common names are used for organic compounds, the systematic name is given as well at the first mention. Common names are widely used in the chemical literature, in industry and commerce, and there is a great divergence in the use of systematic as opposed to non-systematic nomenclature in the English-speaking world.

I thank many colleagues for helpful comments and advice, particularly Mr Martyn Berry and Professor Alwyn Davies FRS who have read the entire manuscript and whose suggestions for changes have improved the text in numerous places. I would also like to thank my wife Pat for her support and forbearance over the past three years. Enjoy the book!

Richard A. Jackson
University of Sussex

TUTORIAL CHEMISTRY TEXTS

EDITOR-IN-CHIEF

Professor E W Abel

EXECUTIVE EDITORS

Professor A G Davies
Professor D Phillips
Professor J D Woollins

EDUCATIONAL CONSULTANT

Mr M Berry

This series of books consists of short, single-topic or modular texts, concentrating on the fundamental areas of chemistry taught in undergraduate science courses. Each book provides a concise account of the basic principles underlying a given subject, embodying an independent-learning philosophy and including worked examples. The one topic, one book approach ensures that the series is adaptable to chemistry courses across a variety of institutions.

TITLES IN THE SERIES

Stereochemistry *D G Morris*
Reactions and Characterization of Solids
  S E Dann
Main Group Chemistry *W Henderson*
d- and f-Block Chemistry *C J Jones*
Structure and Bonding *J Barrett*
Functional Group Chemistry *J R Hanson*
Organotransition Metal Chemistry *A F Hill*
Heterocyclic Chemistry *M Sainsbury*
Atomic Structure and Periodicity *J Barrett*
Thermodynamics and Statistical Mechanics
  J M Seddon and J D Gale
Basic Atomic and Molecular Spectroscopy
  J M Hollas
Organic Synthetic Methods *J R Hanson*
Aromatic Chemistry *J D Hepworth,*
  D R Waring and M J Waring
Quantum Mechanics for Chemists
  D O Hayward
Peptides and Proteins *S Doonan*
Biophysical Chemistry *A Cooper*
Natural Products: The Secondary
  Metabolites *J R Hanson*
Maths for Chemists, Volume I, Numbers,
  Functions and Calculus *M Cockett and*
  G Doggett
Maths for Chemists, Volume II, Power Series,
  Complex Numbers and Linear Algebra
  M Cockett and G Doggett
Nucleic Acids *S Doonan*

TITLES IN THE SERIES

Inorganic Chemistry in Aqueous Solution
  J Barrett
Organic Spectroscopic Analysis
  R J Anderson, D J Bendell and
  P W Groundwater
Mechanisms in Organic Reactions
  R A Jackson

Further information about this series is available at www.rsc.org/tct

Order and enquiries should be sent to:
Sales and Customer Care, Royal Society of Chemistry, Thomas Graham House,
Science Park, Milton Road, Cambridge CB4 0WF, UK

Tel: +44 1223 432360; Fax: +44 1223 426017; Email: sales@rsc.org

Contents

1
What Is a Mechanism?

The chemical structure of most organic compounds is well established. Spectroscopic methods and X-ray crystallography show that individual atoms in a molecule are connected, usually by covalent bonds. Bond lengths are often known to within about ± 1 pm (0.01 Å) and bond angles to within $\pm 1°$. Molecular models and graphics programs give a good picture of the overall shape of the molecule, including possible interactions between atoms that are not covalently bonded to each other. These structures correspond to energy minima.

Molecules can acquire extra energy by collisions, and this energy may cause distortions of bond lengths or angles by small amounts. However, the bond lengths and angles will tend to return to the equilibrium values.

However, if the distortions become too great, one or more covalent bonds may break, and new bonds may be formed, either within the molecule or with a new molecule with which the first has collided. A chemical reaction has occurred, and when equilibrium is reached, one or more new molecules will be produced, which may be stable or may undergo further reactions.

The energy required to break covalent bonds may be provided thermally by molecular collisions, which give a range of molecular energies, providing some molecules with enough energy to react. At higher temperatures, more molecules will have sufficient energy to react, so the reaction will be faster. Alternatively, the energy can be provided in other ways, especially by visible or UV light. Absorption of a photon by a molecule causes electronic excitation, and the excited state of the molecule may then undergo reactions which cannot be carried out thermally.

Aims

This chapter describes the main features of organic reaction mechanisms and how "reasonable" mechanisms can be written. Having worked through this chapter you should be able to:

- Explain the importance of reaction mechanisms in organic chemistry
- Understand the difference between elementary and stepwise reactions, and the role played by transition states and intermediates
- Know the main types of bond-breaking and bond-making processes
- Identify the bonds made and broken in a reaction, given the starting materials and products, thereby allowing a number of possible reaction mechanisms to be written
- Cut down the number of possible mechanisms by using the concepts of energy and molecularity requirements

1.1   Elementary and Stepwise Reactions

Reactions are of two types. In **elementary reactions** the reacting molecule or molecules are transformed into products directly, without the formation of intermediates. In a **stepwise reaction**, one or more intermediate species are produced, which react further to give the products. A stepwise reaction can be split up into two or more elementary reactions.

As an elementary reaction proceeds, the Gibbs free energy increases up to a maximum value and then goes down to a value corresponding to that of the products. The position of highest energy is called the **transition state**, and is a key feature of the reaction; most of the experimental information about chemical reactions relates to the transition state and will be discussed in the next two chapters.

In a stepwise reaction, at least one of the products of the first elementary reaction reacts further in a second elementary reaction. This may be followed by further elementary reactions until the reaction is complete. Any molecules produced in the course of a stepwise reaction which react further and are not present at the end of the reaction are known as **intermediates**. Intermediates are discussed in more detail in Chapters 4, 5 and 6.

Figure 1.1 shows free energy diagrams for an elementary reaction (1.1a) and for a stepwise reaction with two steps (1.1b).

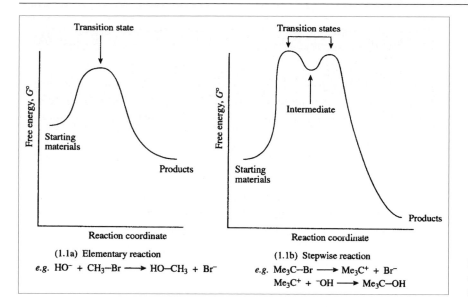

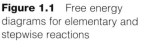

Figure 1.1 Free energy diagrams for elementary and stepwise reactions

An example of an elementary reaction (1.1a) is the S_N2 displacement of a bromide ion from bromomethane by the hydroxide anion. The reaction is thermodynamically favourable (negative $\Delta G°$) and takes place when a hydroxide ion collides with a bromomethane molecule. A bond starts to form between the oxygen atom and the carbon atom at the same time as the carbon-bromine bond breaks. Because of electron repulsion, in the early stages of the reaction the energy released by the formation of the new bond is not quite as much as the energy required to break the C–Br bond, so the free energy increases and eventually reaches a maximum at the transition state, before decreasing to the value appropriate to the products. No intermediate is involved; the reaction proceeds smoothly from reagents through the transition state directly to products.

In contrast, the hydrolysis of *tert*-butyl bromide (2-bromo-2-methylpropane) occurs in a stepwise manner (reaction 1.1b). In the first slow step, the C–Br bond breaks, with the bromine atom taking both electrons from the bond and leaving as a negatively charged bromide ion. The remainder of the molecule is the positively charged *tert*-butyl cation (2-methylprop-2-yl cation). This is a highly reactive intermediate, which reacts rapidly with the hydroxide ion to form the corresponding alcohol.

> **Box 1.1   Elementary and Stepwise Reactions**
>
> In an elementary reaction, reacting molecules are transformed into products without going through intermediates. A stepwise reaction involves consecutive elementary reactions where the intermediate(s) produced in the first elementary reaction react further in subsequent elementary reactions.

Mechanisms cannot be established just by looking at the structures of reactants and products.

These two examples raise an important point about mechanisms. The two reactions are similar from the point of view of reagents and products, yet are known to have different mechanisms. Thus we cannot determine mechanisms merely by knowing the starting materials and products; we need further information. The remainder of this chapter is devoted to writing sensible possible mechanisms for a new reaction, and the remainder of the book concerns the methods that we can use to distinguish between the various mechanistic possibilities.

1.2   Bond Making and Bond Breaking

Inspection of the structure of the reagents and products tells us which covalent bonds have been broken and formed during the reaction.

1.2.1 Bond Breaking

Covalent bonds, which involve two electrons, can be broken in two ways, **homolytically** (homo = same), when one electron is retained by each fragment, or **heterolytically** (hetero = different), when both electrons go to one of the fragments. Heterolysis is more likely if the two atoms in the bond have different electronegativities, and in polar solvents which stabilize charges. Electronegativity is a measure of the power of an atom or a group of atoms to attract electrons from other parts of the same molecule: fluorine is the most electronegative element; cesium is the most electropositive. A covalent bond between two different elements is polarized in the direction $^{\delta+}X–Y^{\delta-}$, where Y is the more electronegative element. In a heterolysis, the bond will almost always break in the direction which will leave both bonding electrons on the more electronegative atom. If the original molecule has no net charge, this will give an anion centred on the more electronegative element and the cation centred on the less electronegative element. The electronegativities of elements commonly found in organic compounds are listed in Table 1.1.

Table 1.1 Electronegativities of elements commonly found in organic compounds[a]

H						
2.1						
Li	Be	B	C	N	O	F
1.0	1.5	2.0	2.6	3.0	3.4	4.0
Na	Mg	Al	Si	P	S	Cl
0.9	1.3	1.6	1.9	2.2	2.6	3.2
K						Br
0.8						3.0
						I
						2.7

[a]Low numbers = electropositive; high numbers = electronegative

The first stage of the S_N1 reaction in reaction (1.1b) is a heterolysis involving a neutral molecule dissociating into a cation and an anion. Heterolysis will take place exclusively in the direction indicated in reaction (1.2a), with no contribution from (1.2b). For the heterolysis shown in (1.2a), the electrons originate in the covalent bond; both move to the bromine atom during the reaction.

$$Me_3C-Br \longrightarrow Me_3C^+ + Br^- \qquad (1.2a)$$

$$(Me_3C-Br \xrightarrow{\quad\times\quad} Me_3C^- + Br^+ \qquad (1.2b)$$

$$Me_3C-O-O-CMe_3 \longrightarrow 2\ Mc_3C-O^\bullet \qquad (1.2c)$$

$$Me_3C^+ \quad ^-OH \longrightarrow Me_3C-OH \qquad (1.2d)$$

$$Cl^\bullet \quad ^\bullet Cl \longrightarrow Cl-Cl \qquad (1.2e)$$

$$HO^- \quad CH_3-Br \longrightarrow HO-CH_3 + Br^- \qquad (1.2f)$$

$$\begin{array}{c} R \\ \diagdown \\ C=O \\ \diagup \\ H \end{array} \quad ^-CN \longrightarrow \begin{array}{c} R \quad CN \\ \diagdown C \diagup \\ \diagup \quad \diagdown \\ H \quad O^- \end{array} \qquad (1.2g)$$

$$HO^- + CH_3-Br \xrightarrow{\quad\times\quad} HO-\underset{\underset{H}{|}}{\overset{\overset{H}{|}}{C}}-Br \longrightarrow HO-CH_3 + Br^- \qquad (1.2h)$$

Homolysis is more likely for weak covalent bonds and for bonds containing atoms with similar electronegativities. Heating di-*tert*-butyl

We show the movement of an **electron pair** in a reaction by a **curved arrow**. The tail of the arrow shows the origin of the electrons that move, the head shows where they end up; thus the arrow gives a good representation of the movement of an electron pair during a reaction.

Half-headed curved arrows (fish-hooks), may be used to show the movement of **single electrons** during a free radical reaction. Because they are more awkward to draw and give a cluttered look to the reaction scheme, curved arrows are less commonly used in homolytic than in heterolytic processes.

peroxide affords an example of homolysis (equation 1.2c). The central O–O bond breaks homolytically to give two *tert*-butoxyl radicals. Since one electron from the two shared electrons in the covalent bond goes to each atom, no charge is created during an homolysis.

1.2.2 Bond Formation

This is the reverse of bond breaking. The electrons involved in the new bond may both come from one of the reagents; this is the reverse of heterolysis. An example is the second step of reaction (1.1b), redrawn in (1.2d) to show the movement of an electron pair from an unshared pair on the oxygen atom (arrow-tail) to a position between the oxygen and the carbon atom where the covalent bond will form (arrow-head). Since an unshared electron pair becomes shared during bond formation, the oxygen effectively loses one electron and the formal negative charge on the oxygen atom is reduced from −1 to zero. Likewise, the positively charged carbon atom acquires a half-share of two electrons and its formal charge changes from + 1 to zero.

If one electron comes from each of the atoms forming the new bond (the reverse of homolysis), there will be no change in formal charge (half of two shared electrons is equivalent to one attached to an individual atom). The combination of two chlorine atoms to form a chlorine molecule is shown in reaction (1.2e); the curved half-arrows show the movement of the electron from the chlorine atom (tail) to the position where the bond will be (head).

Box 1.2   Formal Charge

A carbon atom is electrically neutral if it has four electrons in its valence shell. It is also formally neutral in methane: though surrounded by eight electrons, these are all shared, and are equally involved in balancing the charge on the carbon and the four hydrogen atoms. For each electron pair, one can be formally considered as neutralizing the positive charge on a hydrogen atom; the other will contribute to neutralizing the charge on the carbon atom. Thus the carbon atom will formally have $8/2 = 4$ electrons in its valence shell and will be electrically neutral. In a carbanion such as CH_3^- (see below) there are six shared electrons (counting $6/2 = 3$) and two unshared electrons (belonging exclusively to the carbon atom). This gives five electrons, one more than the number needed for a neutral carbon atom, so the carbon atom has a formal negative charge. Formal charges for some carbon, nitrogen and oxygen species are shown below.

Labelling can sometimes be carried out by introduction of an inert substituent group in the molecule, rather than by isotopic substitution. For example, 4-iodotoluene (4-iodomethylbenzene) can be used to detect the change of position of the substituent group in reaction (1.7). The products are a mixture of 3- and 4-aminotoluenes, but not the 2-isomer, again pointing to the benzyne intermediate, rather than to complete freedom of attack for the incoming amide ion.

1.3   Molecularity

We conclude this chapter by considering the **molecularity** of elementary reactions, that is the number of molecules that are involved in the transition state. This number is almost always one or two. Such reactions are termed **unimolecular** and **bimolecular**.

1.3.1 Unimolecular Reactions

Unimolecular reactions may be concerted, involving simultaneous bond formation and cleavage, as in reaction (1.8), or may involve the breakage of one bond, either heterolytically, as in the first stage of (1.1b), or homolytically, as in (1.4b). The energy required for reaction may be acquired by random collisions, which occasionally give a molecule the energy required for reaction. Photolysis, in which a photon of visible or UV light is absorbed by a molecule, may also cause reaction, often by homolysis of a covalent bond, for example the photolysis of a chlorine molecule to give two chlorine atoms (equation 1.9).

$$\text{(1.8)}$$

$$\text{Cl}-\text{Cl} \xrightarrow{\ h\nu\ } 2\ \text{Cl}^{\bullet} \qquad (1.9)$$

1.3.2 Bimolecular Reactions

Bimolecular reactions involve a collision between two molecules, with enough energy to overcome the activation barrier. These processes are usually concerted, with bond formation and breaking taking place simultaneously. The relative orientation is important, so that the new bonds can be formed between atoms that are near enough to each other. Reactions (1.1a) and the first step of (1.7b) are examples of bimolecular reactions. Reaction (1.10) is an example of a bimolecular reaction that does not involve ions. Three covalent bonds are broken and formed synchronously, and **Diels–Alder reactions** of this type are very useful synthetically, particularly in forming compounds with new six-membered rings.

In reaction (1.10), the electron pairs are shown flowing in the clockwise direction. They could equally well have been shown as flowing in the anticlockwise direction or as single electron movements. See Problem 1.4.

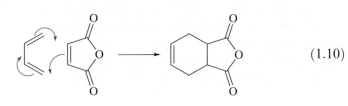

$$(1.10)$$

1.3.3 Termolecular Reactions

Termolecular reactions, involving three molecules in the transition state, are very rarely encountered. Bimolecular reactions have to take place in the very brief time that two molecules collide, before they bounce apart again. The chance that a third molecule will collide at exactly the same time, in a suitable orientation for reaction, is extremely improbable. Termolecular reactions only occur under very unusual conditions, where special circumstances apply. For example, the reaction of two hydrogen atoms in the gas phase to form a hydrogen molecule (reaction 1.11) cannot take place as a bimolecular reaction because the energy liberated by the formation of the H–H covalent bond can only go into vibrational and rotational energy of the new molecule, and this energy is sufficient to cause the almost immediate reversal of the reaction to give back the two hydrogen atoms. However, if a third molecule is available to absorb the energy, reaction (1.11) can take place.

$$H^{\bullet} + H^{\bullet} + M \longrightarrow H{-}H + M \qquad (1.11)$$

Almost all reactions take place in steps that are either unimolecular or bimolecular.

Reactions involving more than three molecules are virtually impossible.

1.3.4 Microscopic Reversibility

In principle, all elementary reactions are reversible, although if they are very exothermic the reverse reaction may be immeasurably slow. The reverse reaction must follow the same route (in reverse) as the forward reaction and go *via* the same transition state. This is known as the **Principle of Microscopic Reversibility**. The main practical application of this principle is in ruling out, as elementary processes, reactions which would give more than three molecules. The reverse of any such reaction would have a molecularity greater than three, which we have established above would be virtually impossible. Mechanism (1.4a) can be ruled out on these grounds, and another example will be found in the problems at the end of the chapter.

1.4 Formulating Mechanisms

We have seen above in reaction (1.1) that there are at least two possible mechanisms for the hydrolysis of halogenoalkanes. How do we formulate possible mechanisms?

First, inspect the bonding in the reagents and products, to find what bonds have been broken and formed. Then, use this information to create possible mechanistic paths based on the different ways in which these bond-breaking and -making processes can be achieved, trying different possibilities involving bond breakage first, bond formation first, or simultaneous breakage and formation of two or more bonds. Remember that bonds can break homolytically or heterolytically.

Rule out steps involving more than one net bond-breaking process taking place at the same time, and bond-forming reactions that would expand the octets of first-row elements on energetic grounds. Rule out steps that involve more than two reactant or three product molecules on "probability of collision grounds" (see Section 1.3). Look for concerted reactions where possible; reactions where bond breakage and formation take place at the same time will usually have lower activation energies than reaction steps involving bond breakage.

If a catalyst is required for the reaction, ensure that it figures in your mechanistic scheme. A bond will need to be established between the catalyst and one of the reacting molecules, usually in the first step of the reaction. In a later step, this bond will be broken again to regenerate the catalyst.

Heterolytic processes are favoured in polar solvents where the ions formed are stabilized, but are uncommon in non-polar solvents and virtually unknown in the gas phase. Homolytic and molecular processes are much less affected by solvents, and are therefore the favoured possibilities for reactions taking place in the gas phase or in non-polar solvents.

Worked Problem 1.2

Q Aromatic aldehydes such as benzaldehyde (phenylmethanal), PhCH=O, react with HCN in water to give the cyanohydrin PhCH(OH)CN. Suggest possible mechanisms for this reaction. *Hint*: HCN is partially dissociated in water, with the formation of H_3O^+ and ^-CN ions.

A The C=O π bond is broken and C–CN and O–H bonds are formed. In route (a), the first step involves nucleophilic addition of the ^-CN ion to the carbonyl carbon, with the simultaneous breakage of the C=O π bond to give the intermediate anion, which in a second step acquires a proton from a hydroxonium ion (or a water molecule) to give the cyanohydrin.

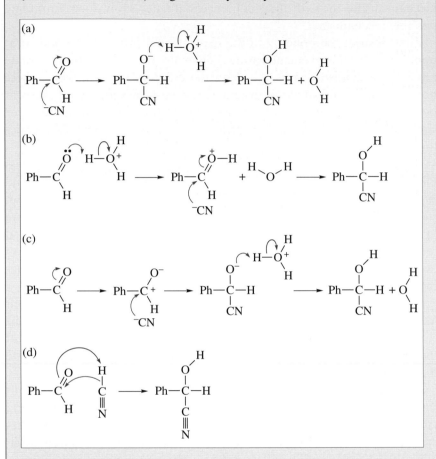

Alternatively, in route (b) the O–H bond may be formed first, followed by attack of ⁻CN at the carbonyl carbon atom. In both these routes, in each of the two steps, a concerted process takes place with the simultaneous formation and breakage of a covalent bond, making these routes likely candidates.

Route (c) involves the unimolecular breakage of the C=O π bond in the first step. This will require energy, which is not offset, as in (a) and (b), by simultaneous formation of a new covalent bond, making (c) less likely. The molecular reaction shown in (d) looks plausible, as two bonds are simultaneously broken and formed. A reason why a mechanism of this type is unlikely is given in Chapter 7.

Worked Problem 1.3

Q   Acetaldehyde (ethanal) reacts in aqueous solution, in the presence of sodium hydroxide, to form 3-hydroxybutanal (aldol), $CH_3CH(OH)CH_2CHO$. The reaction does not take place in the absence of the base. Formulate a possible mechanism or mechanisms.

A   Two molecules of acetaldehyde are needed to give the product. In one molecule a C–H bond is broken, and in the other the π bond of the carbonyl group is broken. Bonds formed are an O–H bond and a C–C bond between the two carbon fragments. The reaction takes place in aqueous solution and the catalyst, sodium hydroxide, is ionic, so an ionic mechanism is likely.

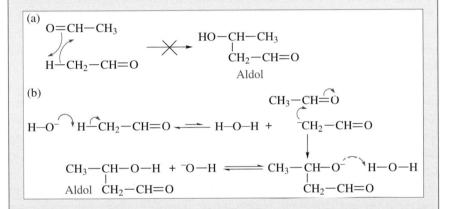

Mechanism (a) does not show the involvement of the hydroxide ion and must therefore be ruled out. Mechanism (b) invokes the removal of a proton from the methyl group of an acetaldehyde molecule by a hydroxide ion to give an unstable carbanion intermediate. This reacts with a second acetaldehyde molecule by nucleophilic addition to the carbonyl group. The resultant anion picks up a proton from a water molecule to give the 3-hydroxybutanal product, and regenerates the hydroxide anion catalyst. All three steps are concerted and involve the simultaneous breakage and formation of covalent bonds. Multi-step concerted reactions of this type are common in biological systems.

1.5   Why Study Mechanisms?

There are a number of reasons, not mutually exclusive, which may be listed by the type of person most affected:

Students. Mechanisms form a framework on which the factual detail of organic chemistry, necessary for a good understanding, can be hung. It would be possible to learn all the individual reagents which add to carbonyl groups, but the classification of many of these as nucleophiles, together with an understanding of why these reagents add to the carbon centre, makes the information more memorable.

Chemists Involved in Synthesis. Mechanistic knowledge allows intelligent variation of reaction conditions, temperatures and proportions of reagents to maximize yields of pure products.

Industrial Chemists. Mechanistic knowledge allows the prediction of new reagents and reaction conditions which may affect desired transformations. It also allows optimization of yields, cutting down on raw materials costs and waste material which may be expensive to dispose of. For example, Augmentin is a broad-spectrum antibiotic, marketed by GlaxoSmithKline, with sales of over $2 billion in 2001. If, say, 5% of the costs are in raw materials, a 1% improvement in reaction yield would save GSK at least a million dollars per annum.

Biochemists, and Those Involved in Medical Research. The reactions involved in metabolism in living organisms are organic, and many are understood in some detail. The establishment of mechanism is of vital importance in understanding how diseases affect metabolism, how drug molecules can assist or prevent particular biochemical reactions, and in the development of new drugs.

Chemists Giving Advice on Environmental Issues. Organic molecules in the environment can have beneficial or harmful effects (or both). An understanding of the mechanistic chemistry involved in the degradation of chemicals in the environment can lead to improvements in the environment. For example, chlorofluorocarbons (CFCs), used as refrigerants, escape into the atmosphere and diffuse to the stratosphere where they damage the ozone layer. Knowledge of the reaction mechanisms involved has led to replacement of these chemicals for some purposes by hydrochlorofluorocarbons (HCFCs). These degrade before they reach the stratosphere and do much less harm to the ozone layer.

Summary of Key Points

1.   *Elementary and Stepwise Reactions*

In an elementary reaction, there are no intermediates. Reagents are converted into products via a transition state in a single step.

A stepwise reaction involves more than one elementary reaction. Intermediates are produced which react further to give products, often involving many steps.

2. *Molecularity*

Elementary steps are normally either unimolecular or bimolecular. In either case, no more than three product molecules are formed.

3. *Formulating Mechanisms*

Note which bonds are made and broken in going from reagents to products, and use these to formulate several possible mechanisms. If a catalyst is needed, it must be involved in the mechanism. Avoid steps which would:

(a) have a molecularity greater than two, or which give more than three product molecules;
(b) produce intermediates with expanded octets of first-period elements;
(c) involve overall breakage (bonds broken *minus* bonds formed) of more than one bond;
(d) require bonds to be formed between atoms that cannot approach each other closely in the transition state;
(e) require the formation of ions in the gas phase or in non-polar solvents.

4. *Determination of Mechanism*

Having formulated several possible mechanisms, try to rule out all but one by focusing on experiments which provide information about transition states (Chapters 2 and 3) or on intermediates (the rest of the book). A mechanism cannot be "proved", but becomes accepted if it is the only one that fits all known experimental facts. A currently accepted mechanism must be modified if new conflicting experimental evidence comes to light.

Further Reading

J. McMurry, *Organic Chemistry*, Brooks/Cole, Pacific Grove, California, USA, 1996, Chapter 5.
R. B. Grossman, *The Art of Writing Reasonable Organic Reaction Mechanisms*, Springer, New York, 1999.

Problems

1.1.   Dibenzylmercury, $(PhCH_2)_2Hg$, decomposes in octane at 140 °C to give 1,2-diphenylethane ($PhCH_2CH_2Ph$) and mercury. Suggest at least two plausible mechanisms for this reaction.

1.2.   Bromine reacts with benzene in the presence of aluminium tribromide to give bromobenzene and hydrogen bromide. Suggest at least two plausible mechanisms for this reaction. *Hint*: Aluminium tribromide and bromine form a complex which can be regarded as $[Br]^+[AlBr_4]^-$.

1.3.   Comment on the following reaction schemes. Each of the processes indicated by an arrow should be considered as a possible elementary reaction.

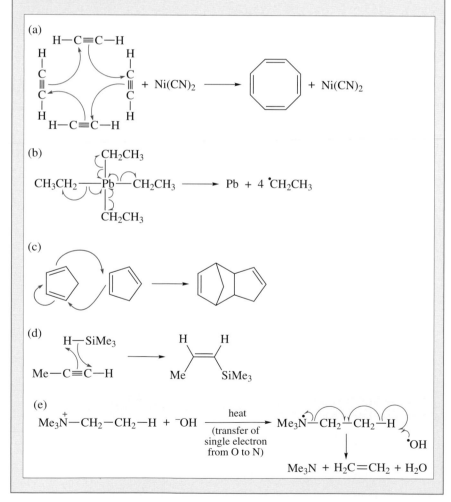

1.4. Reaction (1.10) shows a concerted reaction with three pairs of electrons flowing in a clockwise direction. Show alternative formulations where (a) the electron pairs are shown flowing in the anticlockwise direction, and (b) six single electrons are moving, three clockwise and three anticlockwise.

1.5. Carbene, CH_2, whose electronic structure is shown below, is a reactive intermediate formed in some reactions. Assign a formal charge to the central carbon atom, showing your reasoning.

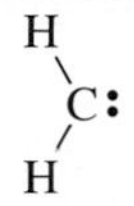

$$H\!\!\diagdown\!\!\underset{H\diagup}{C}\!:$$

2
Kinetics

In Chapter 1, we established some ground rules for writing plausible mechanisms (normally several) for particular reactions, based on the identification of bonds formed and broken in the reaction. In this chapter, we show how **kinetics**, the study of how concentrations of reagents or products vary with time, enable us to rule out some potential mechanisms and provide insight into elementary and stepwise reactions.

Aims

This chapter describes the use of kinetics in supporting or disproving potential mechanisms for a reaction. By the end of this chapter you should be able to:

- Establish zero-, first- and second-order behaviour for particular reaction components from concentration/time data
- Use information about order to support or rule out suggested mechanisms
- Interpret mixed and fractional order kinetic behaviour
- Make inferences from observed Arrhenius parameters in simple cases
- Make inferences from primary kinetic isotope effects

2.1 Rates and Rate Constants

2.1.1 First-order Reactions

When we study the rate of a reaction, we normally actually measure concentrations of reagents or products at different times as the reaction proceeds. To see how this is connected to mechanism, let us look first at elementary unimolecular and bimolecular reactions.

Kinetic theory tells us that reactions take place because random collisions between molecules produce a small number of molecules with an energy greater than the minimum (the activation energy) for reaction to occur. For unimolecular reactions at a particular temperature, this number (and thus the rate of reaction) will be proportional to the number of molecules present in a particular space (volume), *i.e.* the concentration. Thus for a reaction

$$A \rightarrow B \qquad (2.1)$$

$$\text{Rate of reaction} = d[B]/dt = -d[A]/dt = k_1[A] \qquad (2.2)$$

The constant k_1 is known as the **first-order rate constant** and has the units of 1/time, normally expressed as s^{-1}.

For a reaction of this type, the reaction will slow down as the reaction proceeds, so that when half the starting compound has been used up, the rate will have fallen to half of the original value. After three-quarters has been used up (and only a quarter remains), the rate will have fallen to a quarter of the original rate, and so on. The time taken for the concentration to drop to a half of the original value in a first-order reaction is a constant, the **half-life**; this does not depend on the original concentration.

It is easier to measure concentration than rate; a plot of concentration against time is shown in Figure 2.1(a). This type of process is known as exponential decay. In principle, reactions of this type are never complete. However, 99.9% completion corresponds to about 10 half-lives, and 99.9999% to 20 half-lives, at which point for all practical purposes the reaction is complete.

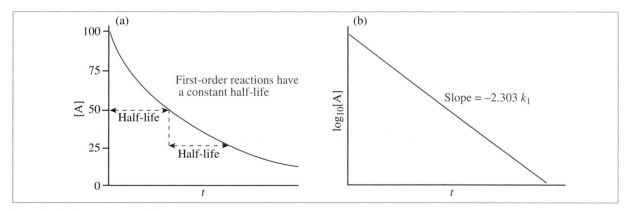

Figure 2.1 First-order reactions

To find out if the reaction is first order, you can make a plot of the type shown in Figure 2.1(a) and check that the half-life is constant, *i.e.* that the time taken for the concentration to fall from 100% to 50% is the same as that for 50% to 25% and 25% to 12.5%. In practice, it is often difficult to be sure of half-lives, particularly if there is significant scatter, and it may be impractical to measure the rate over several half-lives.

An alternative is to use calculus to transform equation (2.2) by integration. This gives equations (2.3) and (2.4):

$$\ln\{[A]_0/[A]_t\} = k_1 t \tag{2.3}$$

$$\log_{10}[A] = -2.303k_1 t + \log_{10}[A]_0 \tag{2.4}$$

Thus a plot of $\log_{10}[A]$ against time should give a straight line of slope $-2.303k_1$. The straightness of the plot is evidence for first-order behaviour, and the slope allows the first-order rate constant k_1 to be determined (Figure 2.1b).

From a practical point of view, it is often easier to monitor the concentration of product rather than reagent. Provided that the reaction is quantitative, the final concentration of B, $[B]_\infty$, will equal $[A]_0$, so a plot of $[B]_\infty - [B]_t$ against t will be equivalent to a plot of $[A]_t$ against t.

2.1.2 Second-order Reactions

For an elementary bimolecular reaction, the two molecules involved may be the same or different. The number of product molecules is almost always one or two. We have already come across reactions of this type: the S_N2 hydrolysis of halogenoalkanes involves reaction between two different molecules; reaction (2.5) involves two molecules of the same compound, buta-1,3-diene (**1**):

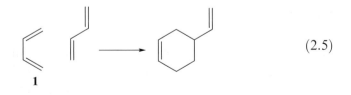

$$\tag{2.5}$$

1

Since two molecules are involved in the reaction, and the rate of reaction depends on the frequency of collisions, the rate of reaction will depend on both [A] and [B] (different reagents), or on $[A]^2$ if there is only one reagent:

$$A + B \longrightarrow C + D \tag{2.6}$$

$$\text{Rate of reaction} = d[C]/dt = d[D]/dt = -d[A]/dt$$
$$= -d[B]/dt = k_2[A][B] \tag{2.7}$$

$$A + A \longrightarrow C \tag{2.8}$$

$$\text{Rate of reaction} = -d[A]/dt = k_2[A]^2 \tag{2.9}$$

For reactions involving a single component A, a plot of [A] against time will give a curve of the type shown in Figure 2.2(a). Because the rate of reaction depends on the square of the concentration of the reagents, the rate will fall off more rapidly with time than for a first-order reaction, and the half-life will increase as the reaction proceeds. As with first-order reactions, it is useful to integrate the rate equation to give an expression involving concentration directly; the result of integrating equation (2.9) is equation (2.10), from which it can be seen that a plot of $1/[A]$ against time should be a straight line, with a slope of k_2. This is shown in Figure 2.2(b).

$$1/[A] - 1/[A]_0 = k_2 t \tag{2.10}$$

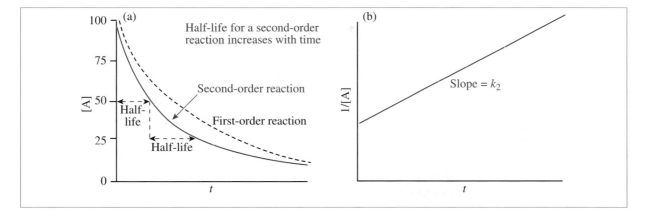

Figure 2.2   Second-order reactions

Similar expressions can be derived for reactions of different orders.

For reactions with two components, it is better to determine the order with respect to each component separately, as described in the next section.

2.1.3 Order of Reaction and Connection between Reaction Order and Reaction Mechanism

In more complicated reactions with several reagents, the reaction rate depends on the concentrations of some or all of the components, and for a reaction of type (2.11), an expression of the type (2.12) can often be written:

$$A + B + C \ldots \longrightarrow Products \qquad (2.11)$$

$$Rate = -d[A]/dt = -d[B]/dt = -d[C]/dt = k[A]^{l}[B]^{m}[C]^{n} \ldots \quad (2.12)$$

The reaction is then said to be of l-th order with respect to A, m-th order with respect to B, n-th order with respect to C, and so on, with a total order of $l + m + n$. The orders with respect to each component are usually but not always integers. Experimentally, orders with respect to individual reagents are usually obtained by carrying out the reaction with all other components except the one being investigated (say A) being in a large (10- to 20-fold) excess. Under these conditions, the concentrations of B, C, *etc.*, will not change appreciably during the reaction, and the rate will effectively depend only on [A] as the only reagent being depleted. We talk of pseudo-first-order, pseudo-second-order, *etc.*, behaviour in these cases.

Worked Problem 2.1

Q   The following data were obtained for the disappearance of a reagent X in a reaction in which all other components are present in a 20-fold excess. Use appropriate plots to determine, if possible, if the reaction is zero, first, or second order with respect to the reagent concerned.

t/s	0	60	120	180	240	300
[X]/M	1.0	0.66	0.43	0.29	0.18	0.10

A   Plots of [X], $\log_{10}[X]$ and $1/[X]$ against time are conveniently obtained using a spreadsheet and are shown below. As outlined in Section 2.1, these plots should produce straight lines for zero-, first- or second-order dependence respectively of the rate with respect to component X. Inspection of the plot shows that curve (b) corresponding to first-order behaviour is straight, whereas the other two plots are curved. These plots provide convincing evidence of first-order behaviour with respect to X, provided that the order is an integer.

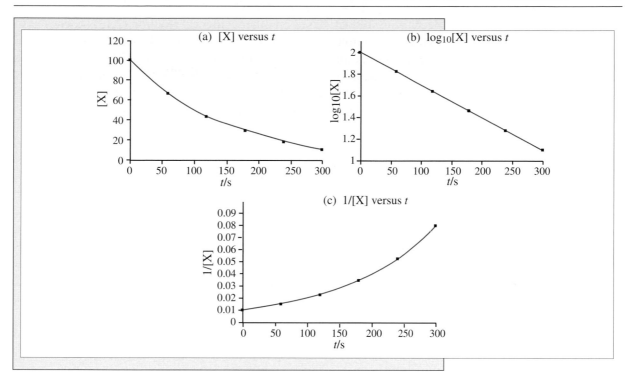

2.2   Conclusions about Mechanism that can be Drawn from Kinetic Order

2.2.1   First- and Second-order Reactions

Elementary uni- and bimolecular reactions will necessarily show first- and second-order kinetic behaviour, but the reverse is not necessarily true: a first-order reaction *may* not be unimolecular and a second-order reaction *may* not be bimolecular. For example, we considered the decomposition of dibenzylmercury in Chapter 1, in which the mechanism could either be elementary, giving a mercury atom and a 1,2-diphenylethane molecule directly (reaction 2.13a), or the reaction could be complex, with a slow initial homolysis of a carbon–mercury bond, followed by rapid further reactions to give the products (reaction 2.13b). Similarly for the Cope rearrangement of diene **2** to diene **4**, the reaction could be elementary, with a concerted cyclic movement of electrons (reaction 2.14a), or might involve a di-radical intermediate **3** which rapidly reacted further to give the observed product **4** (reaction 2.14b). Both these mechanisms would lead to first-order kinetics, so the establishment of first-order kinetic behaviour for both these reaction schemes does not establish the

Elementary uni- and bimolecular reactions must show first- and second-order kinetics, respectively; however, reactions showing first- and second-order kinetics are not necessarily elementary.

mechanism. In fact, based on other evidence, reaction (2.13) is believed to involve the complex mechanism (2.13b) whereas the Cope rearrangement is believed to be an elementary unimolecular reaction (2.14a). The same is true for reactions between two different species to give products: second-order rate dependence (first order with respect to both components) is consistent with a bimolecular reaction but does not prove it.

$$PhCH_2-Hg-CH_2Ph \longrightarrow PhCH_2-CH_2Ph + Hg \qquad (2.13a)$$

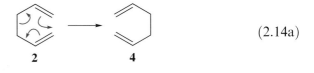

$$(2.13b)$$

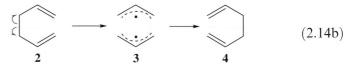

$$(2.14a)$$

$$(2.14b)$$

2.2.2 Rate-determining Step

If a reaction has a rate-determining step, only species that are involved in steps up to and including the rate-determining step can affect the rate of reaction.

Why does the mechanism shown in reaction (2.13b) lead to simple first-order kinetics, not involving rate constants for the reactions subsequent to the first step? The answer is that the first step is *slow*; subsequent reactions are much faster. The only reaction by which dibenzylmercury is destroyed is the first step, which is a unimolecular reaction that is first order. The overall rate of loss of dibenzylmercury will therefore show first-order kinetics, with the measured rate constant corresponding to k_d. Reactions of this type with a "bottleneck" are said to have a **rate-determining** (or **rate-limiting**) **step**. Reactions subsequent to the rate-determining step, which need not be the first step, can have no effect on the kinetics. We have already come across another reaction involving a rate-determining step, the S_N1 hydrolysis of a halogenoalkane by a base. Although reaction (2.15) involves both a *t*-butyl bromide (2-bromo-2-methylpropane) molecule **5** and a hydroxide ion, experimentally the rate shows a first-order dependence on **5** and zero-order (*i.e.* no) dependence on the hydroxide ion, in accordance with equation (2.15). *Reactions which show a zero-order dependence on one of the components cannot be elementary reactions.*

$$Me_3C—Br \xrightarrow{\text{slow}} Br^- + Me_3C^+ \xrightarrow[\text{fast}]{^-OH} Me_3C—OH \qquad (2.15)$$
$$\underset{\textbf{5}}{}$$

Worked Problem 2.2

Q In the acid-catalysed bromination of acetone (propanone) in water, the reaction was followed by measuring the disappearance of bromine by taking aliquots at intervals, adding excess potassium iodide (which converts the remaining bromine into iodine) and titrating the iodine liberated against 0.025 M sodium thiosulfate ($Na_2S_2O_3$) solution. The data shown below were obtained (the $Na_2S_2O_3$ titre is proportional to the residual bromine concentration). What conclusions about the reaction can be drawn?

$$CH_3COCH_3 + Br_2 \xrightarrow{H^+} CH_3COCH_2Br + HBr$$

t/min	0	10	20	60	67	70	75	80	83
0.025 M $Na_2S_2O_3$ solution/mL	19.25	17.05	14.8	5.0	3.5	3.0	1.6	0.65	0.05

A The plot of the $Na_2S_2O_3$ titre, which is proportional to $[Br_2]$, against time shown below in (a) is linear, whereas a logarithmic plot (b) is curved. This shows that the rate of the bromination reaction does not depend on $[Br_2]$. Thus a bromine molecule is not involved in the rate-determining step, so the initial reaction must involve the other components. A possible scheme, consistent with this evidence, is shown below. There will be a zero-order dependence on $[Br_2]$ provided that the rate of reaction c (the only step which involves a bromine molecule) is considerably faster than $-b$. This mechanism would need to be tested further by other experiments, the first of which would be to confirm the kinetic dependence on the other two components.

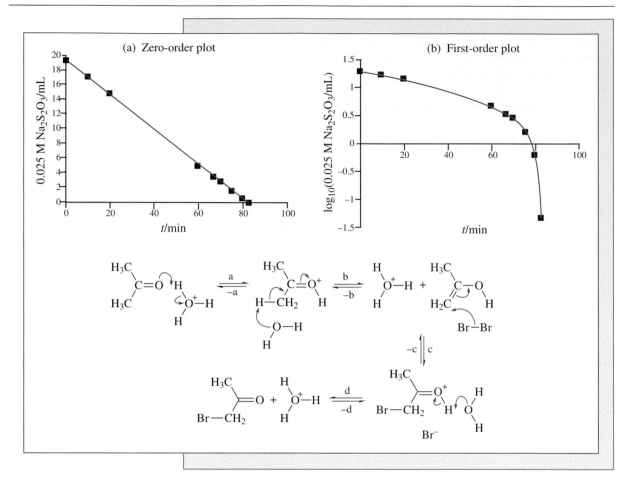

2.2.3 Rate of Formation of Products

Incidentally, if we tried to measure the rates of reactions (2.13) or (2.15) by measuring the rate of appearance of the products, we would find that the rates would almost exactly match those of the disappearance of the starting materials. Even though intermediates are involved, they are so unstable as not to build up any measurable concentration during the reaction. However, in the acid-catalysed hydrolysis of the diester **6** (reaction 2.16), the initial rate of disappearance of **6** will be greater than the rate of formation of **8** because the intermediate **7** will have a similar reactivity to the initial diester **6**, so it will build up in concentration initially, and until an appreciable amount of the intermediate is formed, the rate of production of the diacid product will be negligible. *Reactions where the rate of formation of the products is initially slower than the rate of disappearance of starting materials cannot be elementary.*

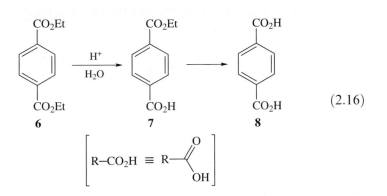

(2.16)

2.2.4 Reactions of Order Higher than Second

These cannot be elementary, since termolecular organic reactions are virtually non-existent. The most common reason for third-order kinetics is for two reagents to react with each other rapidly and reversibly to give an intermediate, which then undergoes a slow rate-determining reaction to give a product. An example is the base-catalysed dimerization of two acetaldehyde (ethanal) molecules to give 3-hydroxypropanal (aldol, **11**). The rate is proportional to $[CH_3CHO]^2[OH^-]$. Proton abstraction by the hydroxide ion from the acetaldehyde molecule gives rise to a small concentration of the intermediate anion **9**. The reverse reaction is rapid, an equilibrium is set up and the concentration of this intermediate, $[9]$, will be $k_{2.17a}[CH_3CHO][OH^-]/k_{-2.17a}[H_2O]$. The rate-determining step is equation (2.17b), a second-order reaction involving addition of **9** to a CH_3CHO molecule to give the anion **10**. The rate of formation of product will therefore be the rate of this reaction, $k_{2.17b}[CH_3CHO] \times [^-CH_2CHO] = k_{2.17a}k_{2.17b}[CH_3CHO]^2[OH^-]/k_{-2.17a}[H_2O]$, accounting for the third-order behaviour.

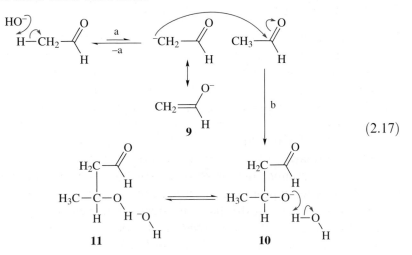

(2.17)

2.2.5 Reactions with Fractional Order: Radical Chain Reactions

These cannot be elementary. The fractions are usually 1/2 or 3/2 and indicate a **radical chain reaction**. An example is the bromination of trichloromethane (chloroform; reaction 2.18); the kinetic expression is given by equation (2.19).

$$Br_2 + CHCl_3 \longrightarrow BrCCl_3 + HBr \qquad (2.18)$$

$$\frac{-d[Br_2]}{dt} = k_{obs}[CHCl_3][Br_2]^{0.5} \qquad (2.19)$$

Radical reactions will be discussed in more detail in Chapter 6. Radicals are molecules or atoms with an unpaired electron, usually formed by thermolysis (*e.g.* reaction 2.20) or by photolysis of compounds containing a weak covalent linkage. In reaction (2.18), the Br–Br bond is the weakest bond present in the two molecules, and the homolysis (2.20) takes place at an appreciable rate at temperatures above 150°C. Processes that produce radicals in this way are termed **initiation reactions**. Since the radicals are formed in very small quantities, and since they are very reactive species, the bromine atom will react predominantly with trichloro-methane (reaction 2.21) to give hydrogen bromide and the trichloro-methyl radical. This in turn reacts with a bromine molecule (usually not a bromine atom – there are very few of these around) to give the bromotrichloromethane product and regenerating a bromine atom (reaction 2.22). The pair of reactions (2.21) and (2.22) are called **propagation reactions**. These are the reactions that turn reagents into products: up to several thousand molecules for each bromine atom produced. The bromine atom used up in (2.21) is regenerated in (2.22). Finally, the combination reactions (2.23), (2.24) and (2.25) destroy the radicals (**termination**). These bimolecular reactions have very large rate constants, so the radical concentrations can only reach a very low level.

$$Br_2 \longrightarrow 2\ Br^{\bullet} \qquad \text{Initiation} \qquad (2.20)$$

$$Br^{\bullet} + H{-}CCl_3 \longrightarrow Br{-}H + {}^{\bullet}CCl_3 \ \Big\} \ \text{Propagation} \qquad (2.21)$$
$$Br{-}Br + {}^{\bullet}CCl_3 \longrightarrow Br^{\bullet} + Br{-}CCl_3 \qquad (2.22)$$

$$Br^{\bullet} + Br^{\bullet} \longrightarrow Br_2 \qquad (2.23)$$
$$Br^{\bullet} + {}^{\bullet}CCl_3 \longrightarrow Br{-}CCl_3 \ \Big\} \ \text{Termination} \qquad (2.24)$$
$$2\ {}^{\bullet}CCl_3 \longrightarrow Cl_3C{-}CCl_3 \qquad (2.25)$$

For this reaction, it turns out that the propagation step (2.21) is much slower than (2.22), so that when a bromine atom is lost in reaction (2.21), it is quickly regenerated by (2.22); therefore most of the radicals present in the system at any one time are bromine atoms. Since the instantaneous concentration ($\sim 10^{-8}$ M) of bromine atoms is much smaller than the throughput of the atoms and its absolute value falls only very slowly as the reaction proceeds, the rates of formation ($= 2k_{20}[Br_2]$) and destruction ($= 2k_{23}[Br^\bullet]$) of these reactive atoms are approximately equal, so we can equate these rates in equation (2.26) and derive an expression for the concentration of the bromine atoms in equation (2.27). This is the **steady-state approximation**. Because $[Br^\bullet] \gg [^\bullet CCl_3]$, termination steps (2.24) and (2.25) can be ignored. Throughput to products depends on the propagation steps, and since (2.21) is rate determining, the overall rate of reaction is given by equation (2.28), and the reaction has an overall order of 1.5: first order with respect to the trichloromethane and 0.5 order with respect to the bromine.

Fractional orders, particularly if half powers are involved, suggest a free radical reaction.

$$\frac{d[Br^\bullet]}{dt} = 2k_{2.20}[Br_2] - 2k_{2.23}[Br^\bullet]^2 \approx 0 \qquad (2.26)$$

$$[Br^\bullet] = (k_{2.20}/k_{2.23})^{0.5}[Br_2]^{0.5} \qquad (2.27)$$

$$\frac{d[BrCCl_3]}{dt} = \frac{-d[CHCl_3]}{dt} = k_{2.21}[CHCl_3][Br^\bullet]$$
$$= k_{2.21}\left(\frac{k_{2.20}}{k_{2.23}}\right)^{0.5}[CHCl_3][Br_2]^{0.5} \qquad (2.28)$$

2.2.6 Reactions with Mixed Order: Competing Reactions

In many reactions, two or more processes may contribute to the disappearance of a reactant. These processes may have different kinetic dependence on the reagent concentrations. For example, when 2-bromopropane is hydrolysed to propan-2-ol by sodium hydroxide in aqueous ethanol, both S_N1 and S_N2 processes take place at the same time, and each contributes to the loss of the 2-bromopropane, as shown in reactions (2.29) and (2.30). Both processes involve $[Me_2CH-Br]$, but the S_N2 process also depends on $[OH^-]$ whereas the S_N1 process does not. The overall rate of loss of 2-bromopropane is given by expression (2.31), showing a mixed kinetic dependence of zero and first order with respect to the hydroxide ion.

$$Me_2CH-Br \xrightarrow{\text{slow}} Br^- + Me_2CH^+ \xrightarrow{^-OH} Me_2CH-OH \qquad (2.29)$$

$$Me_2CH-Br + {}^-OH \longrightarrow Me_2CH-OH + Br^- \qquad (2.30)$$

$$\text{Rate} = \frac{-d[Me_2CHBr]}{dt}$$
$$= k_{2.29}[Me_2CHBr] + k_{2.30}[Me_2CHBr][OH^-] \qquad (2.31)$$

This kinetic behaviour is difficult to distinguish from the fractional order dependence discussed in the previous section. However, if several experiments are carried out with different $[OH^-]$ concentrations, and a plot of {initial rate of reaction/$[Me_2CH-Br]$} against $[OH^-]$ is made, the result will be a straight line of slope k_2 and an intercept of k_1, as shown in Figure 2.3. For this reaction, $k_1 = k_{2.29}$ and $k_2 = k_{2.30}$.

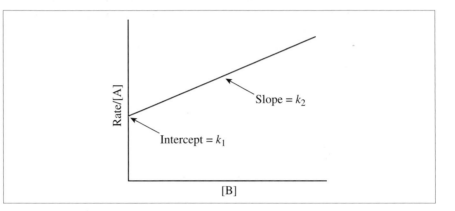

Figure 2.3 Mixed first- and second-order kinetics. A plot of rate/[A] against [B] for a reaction which follows the rate expression: rate = $k_1[A]+k_2[A][B]$

Mixed order kinetics indicate that two or more reactions are taking place in parallel.

Many acid- and base-catalysed reactions show mixed-order kinetics. For example, 1,1,1-triethoxyethane (ethyl orthoacetate, **12**) is hydrolysed to ethanol and acetic acid (ethanoic acid) in aqueous buffers of *m*-nitrophenol (a weak acid) and its sodium salt (reaction 2.32). The kinetic dependence is shown in equation (2.33). It appears that three separate processes contribute to the loss of **12** from the reaction mixture. The second term in rate expression (2.33) suggests a reaction with a rate-determining step involving transfer of a proton from H_3O^+ with a rate constant k_h. This would correspond to the first step of (2.32), with $HA = H_3O^+$. The third term suggests a similar process, but with the transfer of the proton from the undissociated *m*-nitrophenol (first step in 2.32, but with $HA = HOC_6H_4NO_2$, rate constant k_a). The first term probably relates to the transfer of a proton to **12** from a water molecule (first step in 2.32, but with $HA = H_2O$). The k_u term should involve the water concentration, but this cannot be varied significantly in aqueous solution. The mixed order kinetic expression for reaction (2.32) provides strong evidence that the reaction can be catalysed by any acid present (H_3O^+, $HOC_6H_4NO_2$ or H_2O). Any acid can provide the proton in the rate-determining step, though with different rate constants, and the reaction is therefore subject to general acid catalysis.

$$\underset{\textbf{12}}{\text{CH}_3-\overset{\displaystyle \overset{\text{EtO}}{|}}{\underset{\displaystyle \underset{\text{EtO}}{|}}{\text{C}}}-\overset{\diagdown}{\underset{\diagup}{\text{O}}}\overset{\text{Et}}{}} \xrightarrow[\text{slow}]{\text{HA}} \text{CH}_3-\overset{\displaystyle \overset{\text{EtO}}{|}}{\underset{\displaystyle \underset{\text{EtO}}{|}}{\text{C}}}-\overset{\diagup}{\underset{\diagdown}{\text{O}}}\overset{+}{}\overset{\text{H}}{\underset{\text{Et}}{}} \xrightarrow[\text{several steps}]{\text{fast}} \text{CH}_3\text{CO}_2\text{H} + 3\ \text{EtOH} \quad (2.32)$$

$$\frac{-\text{d}[\textbf{12}]}{\text{d}t} = k_u[\textbf{12}] + k_h[\textbf{12}][\text{H}_3\text{O}^+] + k_a[\textbf{12}][\text{HOC}_6\text{H}_4\text{NO}_2] \quad (2.33)$$

2.2.7 More Complex Kinetic Behaviour

Not all reactions show a kinetic behaviour based on a simple order dependence. The most common reason for this is that the reaction is reversible, so the rate will fall as product accumulates, and its concentration will enter the kinetic rate expression. The simplest solution if this happens is to measure initial rates and confine the kinetic study to the first few percent of the reaction.

2.2.8 Summary

A flow chart showing how information about kinetic order gives useful information about mechanisms is shown in Figure 2.4. It must be emphasized that kinetic information can rule out possible mechanisms

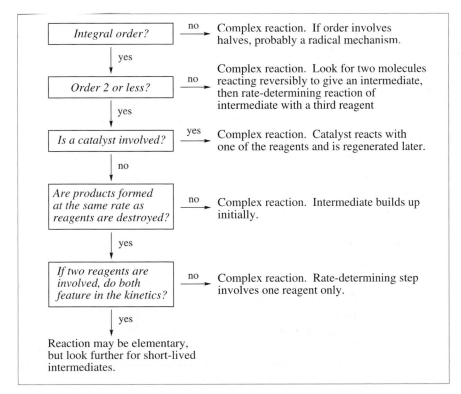

Figure 2.4 Information about reaction mechanisms from kinetics studies

but can never prove a particular mechanism. If more than one mechanistic possibility remains that is consistent with the observed kinetics, then other methods need to be employed, as discussed in the next four chapters.

2.3   The Dependence of Rate of Reaction on Temperature: Arrhenius Parameters

Most reactions go faster at higher temperatures, and for the majority of reactions the variation with temperature can be expressed in the form of equation (2.34). If we take logarithms of both sides of this equation, we obtain equation (2.35), from which we see that a plot of $\log_{10}k$ against $1/T$ (in Kelvin) should be a straight line of slope $E/2.303R$ and intercept $\log_{10}A$. Equation (2.34) is known as the **Arrhenius equation**. A is known as the **Arrhenius A** or **pre-exponential factor** and E is the **activation energy**, which can be identified for elementary reactions as the energy required for the molecules to reach the transition state and therefore be able to react. These parameters provide useful information, particularly for radical or molecular reactions; for ionic reactions, solvent effects make it difficult to extract useful information.

$$k = Ae^{-E/RT} \qquad (2.34)$$

$$\log_{10}k = \log_{10}A - E/2.303RT \qquad (2.35)$$

2.3.1 Information from Activation Energies

Although it is difficult to predict activation energies, it is easier to measure or predict heats (enthalpies) of formation and hence heats of reaction. For an endothermic reaction, the activation energy must be at least as great as the endothermicity ΔH (Figure 2.5), so if the measured activation energy of reaction is less than the endothermicity of a proposed elementary reaction or an initial rate-determining step, that proposal can be ruled out.

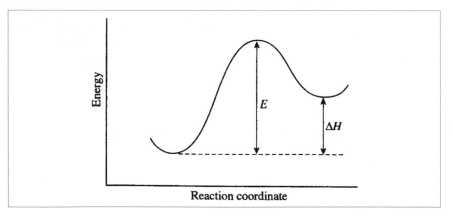

Figure 2.5 Endothermic reactions

For example, tetraethyllead decomposes on heating to give lead, ethane, ethene and butane. Two possible mechanisms are equations (2.36) and (2.37). Mechanism (2.36) involves simultaneous rupture of all four Pb–C bonds to give a lead atom and four ethyl radicals. From the known heats of formation of tetraethyllead, lead and the ethyl radical, the heat of reaction of (2.36) can be estimated as $+515$ kJ mol^{-1}. This is massively greater than the observed activation energy of 155 kJ mol^{-1}, so it must be concluded that (2.36) cannot be correct, which suggests that the stepwise breakage of successive Pb–C bonds as shown in (2.37) is favoured. Since the simple homolysis of one bond in an organic molecule normally takes place without an extra activation energy over and above that required by its endothermicity, the figure of 155 kJ mol^{-1} can be equated with ΔH for the first step of (2.37), allowing the heat of formation of the triethyllead radical to be determined. A large number of heats of formation of organic radicals have been determined in this way.

$$Pb(C_2H_5)_4 \longrightarrow Pb + 4\,C_2H_5^{\bullet} \qquad \Delta H = +515 \text{ kJ mol}^{-1} \qquad (2.36)$$

$$Pb(C_2H_5)_4 \longrightarrow (C_2H_5)_3Pb^{\bullet} + C_2H_5^{\bullet} \xrightarrow{-C_2H_5^{\bullet}} (C_2H_5)_2Pb\!: \xrightarrow{-C_2H_5^{\bullet}} C_2H_5-\overset{\bullet}{Pb}\!:$$

$$\downarrow {-C_2H_5^{\bullet}}$$

$$Pb$$

$$(2.37)$$

Cyclohexene decomposes on heating to give butadiene and ethene. Two mechanistic possibilities are equations (2.38) and (2.39). Mechanism (2.38) involves breakage of a single C–C bond, a process which should require an energy input of about 343 kJ mol^{-1}, lessened by say 38 kJ mol^{-1} to about 305 kJ mol^{-1} to take account of the allylic stabilization expected in the intermediate diradical **13**. The observed activation energy is 279 kJ mol^{-1}, significantly less than this value, which suggests that the alternative concerted reaction (2.39), where three bonds are made concurrently with three bonds being broken, is more likely.

13

$$(2.38)$$

$$(2.39)$$

2.3.2 Arrhenius *A* factors

Low Arrhenius *A* factors suggest a "tight" transition state, with loss of entropy compared with reagents. High *A* factors suggest a "loose" transition state, with a gain of entropy compared with reagents.

Simple kinetic theory for bimolecular reactions equates the activation energy of the reaction to the minimum energy for the reaction to take place, which is brought to the reacting system by the energy of collision of the two reagent molecules. The *A* factor is seen as the product *PZ* of the rate of collisions *Z* multiplied by a probability factor *P*, which depends on the likelihood of a particular collision having the correct geometry for the reaction to take place. Transition state theory relates the *A* factor to the difference in entropy $\Delta S^{\ddagger}$ between the reagent molecule(s) and the transition state. The more entropy that is lost (*e.g.* by molecules associating, forming new bonds that restrict rotation), the lower the *A* factor. This is known as a "tight" transition state. Conversely, if the molecule is dissociating into fragments with a gain in entropy, this corresponds to a "loose" transition state and a high *A* factor. For unimolecular reactions, homolyses of molecules into radicals form excellent examples of a loose transition state, which is close to the two incipient free radicals. These have $\log_{10}A$ in the region of 15–17 s^{-1}. On the other hand, molecular eliminations such as the elimination of HBr from CH_3CH_2Br, which involves a four-membered transition state with some bonding between the H and Br atoms and a restriction of rotation round the C–C bond, have $\log_{10}A$ in the region of 12.5–14 s^{-1}. For bimolecular reactions, the loosest transition states are found in combination reactions of radicals, where (as for the reverse homolysis) the transition state corresponds to very loose association between the two radicals, with most of the rotational entropy conserved. These give $\log_{10}A$ values in the range of 9–10.5 M^{-1} s^{-1}. Radical addition to double bonds and transfer reactions show some loss in entropy in the transition state, and have $\log_{10}A$ values in the region of 7–9 M^{-1} s^{-1}. Cycloadditions such as the Diels–Alder reaction, in which bonds are formed simultaneously between both ends of the reagent molecules, lose even more entropy, with $\log_{10}A$ values in the region of 5–7 M^{-1} s^{-1}. These results are summarized in Table 2.1. It should be emphasized that these generalizations apply only to molecular and radical reactions. For polar reactions, solvation effects are often very large, making generalization difficult.

Table 2.1   Ranges of Arrhenius pre-exponential factors for different reaction types

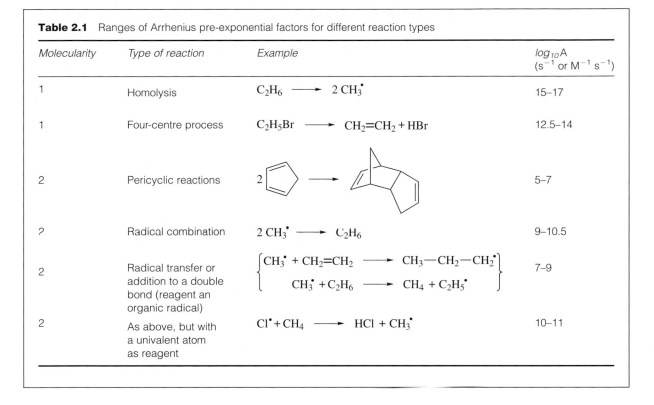

Molecularity	Type of reaction	Example	$log_{10}A$ $(s^{-1}$ or $M^{-1}\,s^{-1})$
1	Homolysis	$C_2H_6 \longrightarrow 2\,CH_3^{\bullet}$	15–17
1	Four-centre process	$C_2H_5Br \longrightarrow CH_2{=}CH_2 + HBr$	12.5–14
2	Pericyclic reactions		5–7
2	Radical combination	$2\,CH_3^{\bullet} \longrightarrow C_2H_6$	9–10.5
2	Radical transfer or addition to a double bond (reagent an organic radical)	$CH_3^{\bullet} + CH_2{=}CH_2 \longrightarrow CH_3{-}CH_2{-}CH_2^{\bullet}$ $CH_3^{\bullet} + C_2H_6 \longrightarrow CH_4 + C_2H_5^{\bullet}$	7–9
2	As above, but with a univalent atom as reagent	$Cl^{\bullet} + CH_4 \longrightarrow HCl + CH_3^{\bullet}$	10–11

2.4   Primary Kinetic Isotope Effects

Isotopes (particularly deuterium) are often used as labels to show where a reaction has occurred in a particular molecule. A specialized use of kinetics applied to isotopically substituted molecules provides information about the mechanism that is often not available from other sources.

The energy of a C–H bond in an organic molecule as a function of bond length is shown in Figure 2.6a. If we compress the bond below its equilibrium value, the energy rises steeply because of electron–electron and nuclear–nuclear repulsion. If we stretch the bond, the energy rises more gradually due to decreased orbital overlap and bonding. At an infinite separation we have the free organic radical R$^{\bullet}$ and a hydrogen atom. The forces involved are electrostatic, and will not be affected if we replace the hydrogen by deuterium, since this merely involves adding an uncharged neutron to the proton in the hydrogen nucleus. However, the deuteron is approximately twice as heavy as the proton, so even though the force constant of the bond is unchanged, the vibration frequency will

Deuterium ($^2H \equiv D$) is a heavy isotope of hydrogen, with a natural abundance 0.02% compared with 99.98% for ^1H. Tritium ($^3H \equiv T$) is a radioactive isotope, obtained from nuclear reactions.

be dramatically reduced because the same force is having to move a particle with twice the mass. At room temperature, most molecules are in their lowest vibrational state, but because of the zero-point energy, this is just above the minimum on the energy curve. This zero-point energy will be slightly smaller for the C–D than the C–H bond, so the deuteriated compound will have a slightly greater bond dissociation energy than the protiated compound, as shown in Figure 2.6a.

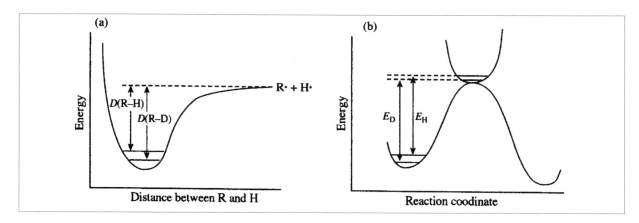

Figure 2.6 The primary deuterium isotope effect. (a) The homolytic dissociation of an alkane, showing the differences between the bond dissociation energies D(R–H) and D(R–D) due to zero-point energy differences (exaggerated). (b) A reaction in which some zero-point energy is retained in the transition state. $E_D - E_H$ is less than D(R–D) – D(R–H)

If the rate-determining step of the reaction involves breakage of the isotopically substituted bond, in the transition state the zero-point energy of that bond will have been lost. Accordingly, there will be a slightly higher activation energy for breakage of the stronger C–D bond than the weaker C–H bond. This is known as the **primary kinetic isotope effect**; k_H/k_D has a maximum value of about 7 at 25 °C and k_H/k_T cannot be greater than about 17 for reactions in which a hydrogen atom is transferred from one species to another. For heavier atoms, the isotope effects are much smaller. The smaller relative difference in isotopic mass ($C_{13}/C_{12} = 1.08$, D/H = 2) will give a much smaller difference in zero-point energies, and the maximum value of a C_{13}/C_{12} isotope effect at room temperature will be in the region of 1.04. The nearer the observed isotope effect is to the theoretical maximum, the more completely the bond in question must be broken in the transition state.

The greater availability and cheapness of deuterium compounds, coupled with the large isotope effects, has until recently made this isotope the most widely used in mechanistic studies. However, recent work using nuclear magnetic resonance or mass spectrometry has allowed the determination of heavy atom isotope effects in natural abundance, by carrying out reactions to a high level of completion and comparing the isotopic composition of the initial reagent with recovered reagent. This technique magnifies small differences in isotope effects into larger differences in ratios of species left after reaction. As an illustration, let us consider two isotopically substituted species A and B present initially in the same concentrations, showing a kinetic isotope effect k_A/k_B of 2. After three half-lives of reaction of the slower species B, its concentration would have fallen to 12.5% of its original value, but the faster species A would have fallen to 1.56%, and the ratio of the two concentrations [B]/[A] in the recovered starting material would be $12.5/1.56 = 8$.

An example of a reaction showing a primary deuterium isotope effect is reaction (2.40), in which acetone (propanone) reacts with bromine in the presence of base to give bromoacetone (bromopropanone). Under similar conditions, the fully deuteriated compound CD_3COCD_3 reacts about seven times more slowly. This is consistent with the mechanism shown; the rate-determining step is the production of the anion **14** in the first step. The C–H bond is partially broken in the transition state, so the deuteriated compound will react more slowly.

$$\text{Me}-\text{CO}-\text{CH}_2-\text{H} \xrightarrow{\text{slow}} \text{Me}-\text{CO}-\text{CH}_2^- \xrightarrow{\text{fast}} \text{Me}-\text{CO}-\text{CH}_2-\text{Br} \quad (2.40)$$

$^-\text{OH} \qquad \textbf{14} \quad \text{Br}-\text{Br} \qquad \qquad \text{Br}^-$

Worked Problem 2.3

Q   Propan-2-ol, Me_2CHOH, is oxidized to acetone (propanone) by chromium(VI) in dilute acid solution by the $HCrO_4^-$ anion at a rate proportional to $[Me_2CHOH][HCrO_4^-][H^+]$. The rate of reaction is reduced by a factor of about 7 if the deuteriated compound Me_2CDOH is used instead. Which of the following mechanisms is consistent with this evidence?

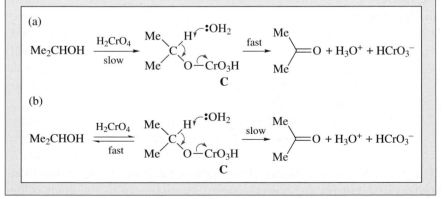

> **A** Mechanism (a) involves a slow, rate-determining step producing the intermediate ester **C**, followed by a fast oxidation step. Since the C–H bond is not broken until after the rate-determining step, there should be no rate change if the propan-2-ol is replaced by Me$_2$CDOH. Thus mechanism (a) can be ruled out.
>
> Mechanism (b) would involve a rapid equilibrium to be set up between the propan-2-ol and its chromate ester **C**. The second step in which the secondary C–H bond is partially broken in the transition state is rate determining, so a substantial isotope effect should be observed. Thus, of the two schemes shown, the substantial isotope effect observed supports (b).

The absence of a primary kinetic isotope effect throws important light on the mechanism of electrophilic aromatic substitution. Nitration of benzene labelled with very small amounts of the radioactive hydrogen isotope tritium was carried out to a partial extent. The radioactivity in the unreacted benzene recovered was virtually identical to that of the initial benzene, showing that there was no significant isotope effect. This suggests that there is no significant breakage of the C–H (or C–T) bond in the transition state, and that equation (2.41b) can be ruled out in favour of (2.41a) where the rate-determining step involves addition of the nitronium ion to give the stabilized intermediate **15**, followed by a rapid loss of a proton (or T$^+$).

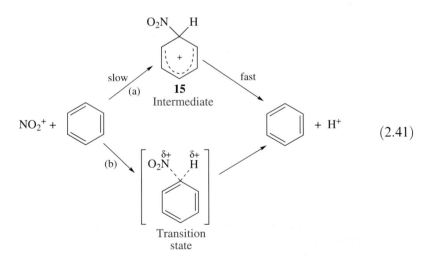

$$(2.41)$$

The primary kinetic isotope effect provides information about bonds broken in steps up to and including the rate-determining step.

The presence of a significant ^{13}C isotope effect at both ends of the diene system (but not for the central carbon atoms) in **16** in the Diels–Alder reaction (2.42) provides strong evidence that both the terminal carbon atoms are changing their bonding in the transition state, supporting the concerted mechanism shown. Likewise, there is a significant ^{13}C isotope effect at both ends of the C=C double bond in the epoxidation reaction (2.43), indicating that both atoms are involved in the transition state, favouring the concerted mechanism shown.

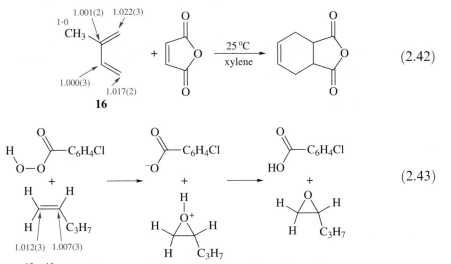

^{12}C/^{13}C kinetic isotope effects at indicated positions; figures in parentheses are standard deviations in the third place of decimals

---

Summary of Key Points

1. Kinetic evidence can cut down the number of mechanistic possibilities for organic reactions, but cannot "prove" a particular mechanism, since more than one mechanism may give the same kinetic order.

2. First- and second-order reactions may be either elementary or stepwise. Third- and higher-order reactions are almost always stepwise.

3. Fractional orders suggest a radical mechanism.

4. Mixed orders suggest two or more competing mechanisms for loss of reagents.

5.   In reactions with a rate-determining step, species that only react *after* the rate-determining step cannot affect the kinetics.

6.   The primary kinetic isotope effect is particularly valuable in showing whether or not a particular bond is involved in the rate-determining step.

Further Reading

B. G. Cox, *Modern Liquid Phase Kinetics*, Oxford University Press, Oxford, 1994.

B. K. Carpenter, *Determination of Organic Reaction Mechanisms*, Wiley, New York, 1984.

N. S. Isaacs, *Physical Organic Chemistry*, Longman, Harlow, 1995, Chapters 2 and 7.

L. Melander and W. H. Saunders, Jr., *Reaction Rates of Isotopic Molecules*, Wiley, New York, 1980.

Problems

2.1.   (a) The following data were obtained for the disappearance of a reagent **A**, in a reaction in which all other components are present in a 20-fold excess. Use plots of [**A**], $\log_{10}$[**A**] and 1/[**A**] against time to determine, if possible, if the reaction is zero, first or second order with respect to **A**.

Time/s	0	20	40	60	80	100
Concentration/M	0.99	0.87	0.75	0.67	0.56	0.51

(b) The same reaction was followed for a further period of time and the following additional data were obtained. Extend your graphs to cover this additional data. Can you refine your conclusions about the order of reaction with respect to **A**?

Time/s	120	140	160	180	200
Concentration/M	0.44	0.39	0.32	0.28	0.22

2.2. Write at least two possible mechanisms which are consistent with each of the following observations:

(a) Bromoethane decomposes in the gas phase to ethene and hydrogen bromide, showing first-order kinetics.

(b) Solutions of bis(trimethylsilyl)mercury, $(Me_3Si)_2Hg$, in inert solvents decompose to give mercury and hexamethyldisilane, $Me_3Si–SiMe_3$, showing second-order kinetics.

2.3. The rate of nitration of benzene in a mixture of nitric acid and sulfuric acid does not depend on the benzene concentration. What can be deduced from this?

2.4. A simplified reaction scheme for the autoxidation of hydrocarbons by oxygen is shown below:

$$\text{Initiator} \xrightarrow{\;k_i\;} 2\,R^{\bullet}$$

$$R^{\bullet} + O{=}O \xrightarrow[\text{fast}]{\;k_{p1}\;} R{-}O{-}O^{\bullet}$$

$$R{-}O{-}O^{\bullet} + H{-}R \xrightarrow[\text{slow}]{\;k_{p2}\;} R{-}O{-}O{-}H + R^{\bullet}$$

$$2\,R{-}O{-}O^{\bullet} \xrightarrow{\;k_t\;} R{-}O{-}O{-}R + O{=}O$$

Oxygen may be presumed to be present in excess. Apply the steady-state treatment to work out the expected kinetics for the reaction. Why is the termination step k_t, which involves two $R{-}O{-}O^{\bullet}$ radicals, likely to be much more significant than alternative termination reactions?

2.5. Suggest a mechanism for the benzidine rearrangement ($\mathbf{B} \longrightarrow \mathbf{C}$) which is consistent with the following information: (1) the reaction is first order with respect to $\mathbf{B}$ and second order with respect to $[H^+]$; (2) the reaction is strictly intramolecular; and (3) there is no isotope effect if the *para*-H atoms are replaced by D.

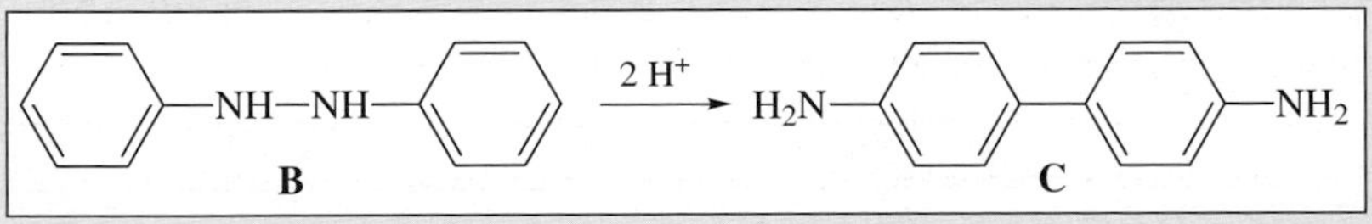

2.6. The Arrhenius A factor for the first-order gas-phase pyrolysis of ethyl acetate is $10^{12.5}$. Which of the two postulated mechanisms shown below does this favour?

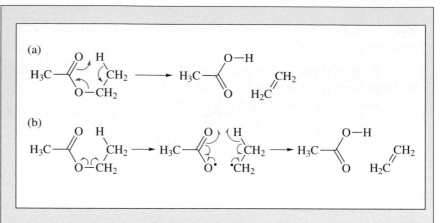

2.7. In the Diels–Alder reaction shown in equation (2.42), rationalize the observation that there is no appreciable ^{13}C isotope effect at the 2- and 3-positions of 2-methylbuta-1,3-diene (**16**).

2.8. In the elimination reaction of the (2-phenylethyl)trimethylammonium ion **D** with the ethoxide ion, there is a ^{14}N/^{15}N nitrogen isotope effect $k_{^{14}N}/k_{^{15}N}$ of 1.0133 ± 0.0002 and a deuterium isotope effect at the position shown (*) k_H/k_D of 3.2. Which of the following mechanisms does this evidence support?

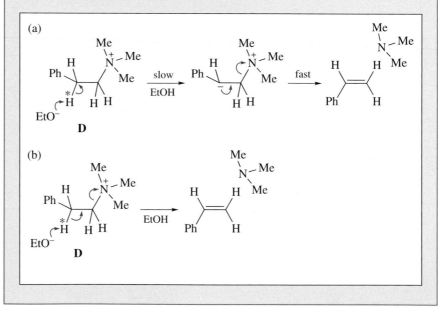

3
The Transition State

━━

We cannot isolate the transition state of a chemical reaction, nor can we study it by spectroscopic methods which rely on the presence of appreciable concentrations of species with particular energy and geometry. However, the transition state does have a geometry and energy; what we can do is to investigate how changes in the reagents and solvents can influence the difference in energy between reagents and the transition state. If these influences make the transition state more stable compared with the reagents, the reaction will be faster; if the transition state is made less stable relative to the reagents, the reaction will be slower. We shall consider the effects of substituents, solvents, steric crowding and stereochemistry. We focus on elementary reactions or rate-determining steps of stepwise reactions.

Aims

By the end of this chapter you should:

- Know how solvents, substituents, steric crowding and stereochemistry can affect the rates of elementary reactions
- Use evidence from these sources to distinguish between or support particular mechanisms
- In the case of substituent effects, be able to identify whether positive or negative charge is being built up in the transition state, and whether the charge is conjugated with the substituent

3.1 Early and Late Transition States

Figure 3.1(a) represents a reaction coordinate reaction for a thermoneutral reaction, for example the abstraction of a hydrogen atom by a methyl radical from a methane molecule (reaction 3.1). Since the products are identical with the reactants, the reaction is exactly thermoneutral, with a transition state involving the transferred hydrogen being equidistant from both methyl carbon atoms. The activation energy of 61 kJ mol^{-1} reflects the fact that because of electron repulsion, each of the partial C–H bonds has less than half the strength of the C–H bond in methane. If we now consider the very similar reaction (3.2), this reaction is exothermic by 25 kJ mol^{-1}, attributed to the stabilization of the ethyl radical compared with the methyl radical. Since the reaction is approximately symmetrical, we may expect the transition state to have approximately half the bonding of the breaking H–CH$_2$CH$_3$ bond and half that of the forming CH$_3$–H bond. Thus we can expect the activation energy to be lowered by about half the change in exothermicity. The observed activation energy of 49 kJ mol^{-1} is in accord with this, and in general we should expect that, for similar approximately thermoneutral reactions, if we change the exothermicity by 2 kJ mol^{-1} we will expect an activation energy change of about 1 kJ mol^{-1}. This is illustrated in Figure 3.1(a).

$$CH_3^{\bullet} \ + \ H–CH_3 \ \longrightarrow \ CH_3–H \ + \ {}^{\bullet}CH_3 \tag{3.1}$$

$$CH_3^{\bullet} \ + \ H–CH_2CH_3 \ \longrightarrow \ CH_3–H \ + \ {}^{\bullet}CH_2CH_3 \tag{3.2}$$

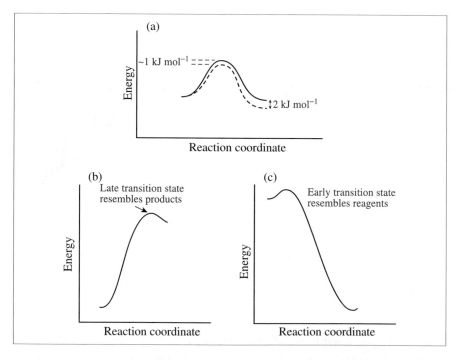

Figure 3.1 The Hammett postulate. (a) Approximately thermoneutral reaction. (b) Very endothermic reaction. (c) Very exothermic reaction

Figure 3.1(b) and (c) show reaction coordinates for very endothermic and exothermic reactions, respectively. The **Hammond postulate** states that for very endothermic reactions the transition state resembles the products almost completely, with a late transition state (Figure 3.1b), whereas for very exothermic reactions the transition state is very like the reagents, *i.e.* an early transition state (Figure 3.1c). There is much experimental evidence to support this view.

For endothermic reactions, since the transition state resembles the products, if a small change is made in the endothermicity of the reaction, almost the whole of the difference will have been developed in the transition state. This will make the rates of reactions of this type very much affected by small changes in endothermicity, often giving rise to strong selectivity between different sites in a molecule for a particular reagent. The reaction that determines the position of bromination of alkanes under free radical conditions is the abstraction of hydrogen by a bromine atom from a C–H bond. This is 44 kJ mol^{-1} endothermic for the reaction with propane to give the prop-1-yl radical (reaction 3.3a), reduced to 29 kJ mol^{-1} for attack at the 2-position (reaction 3.3b). Since these reactions are significantly endothermic, with a late transition state, this will result in a large difference in activation energies for attack at the 1- and 2-positions, giving an increased rate at the 2-position, and hence selectivity in attack there.

$$Br^{\bullet} + C_3H_8 \longrightarrow Br{-}H + {}^{\bullet}CH_2CH_2CH_3 \quad \Delta H = +44 \text{ kJ mol}^{-1} \quad (3.3a)$$
$$\longrightarrow Br{-}H + CH_3\overset{\bullet}{C}H_2CH_3 \quad \Delta H = +29 \text{ kJ mol}^{-1} \quad (3.3b)$$

Conversely, for very exothermic reactions, changes in enthalpy of reaction will have little effect on the activation energy, giving unselective reactions. The corresponding fluorination of propane (reaction 3.4) is highly exothermic: $\Delta H = -159$ kJ mol^{-1} at the 1-position and -173 kJ mol^{-1} at the 2-position. However, with an early transition state and a very small activation energy for both reactions, there is little difference in rate for attack at the two positions and the reaction is very unselective.

Very exothermic reactions have an early transition state and show little positional selectivity. Very endothermic reactions have a late transition state and show high positional selectivity.

$$F^{\bullet} + C_3H_8 \longrightarrow F{-}H + {}^{\bullet}CH_2CH_2CH_3 \quad \Delta H = -159 \text{ kJ mol}^{-1} \quad (3.4a)$$
$$\longrightarrow F{-}H + CH_3\overset{\bullet}{C}H_2CH_3 \quad \Delta H = -173 \text{ kJ mol}^{-1} \quad (3.4b)$$

3.2 Solvent Effects

Ions are stabilized by polar solvents. Hydrogen chloride dissolves in benzene as covalent molecules, but in water it is completely dissociated to give solvated H_3O^+ and Cl^- ions. The high dielectric constant of water decreases the energy of the ions. This principle can be applied to reactions: if charge is being built up in the transition state of a reaction, the transition state will be more stabilized compared with the reagents if the reaction is carried out in a more polar solvent, resulting in an increase in rate. This is shown in Figure 3.2(a). Conversely, if charge is being dispersed or destroyed, the reagents will be more stabilized in more polar solvents, compared with the transition state, resulting in a decrease in reaction rate (Figure 3.2b).

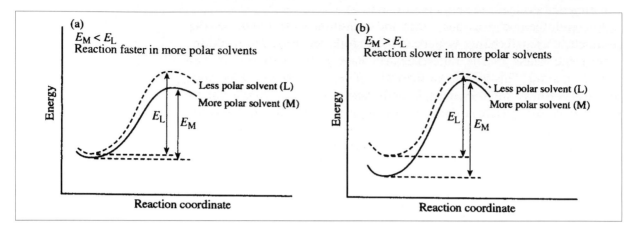

Figure 3.2 Effect of solvent polarity on reaction rate. (a) Charge built up in the transition state. (b) Charge destroyed or dispersed in the transition state

Solvent effects show whether charge is created or destroyed/dispersed in the transition state, but do not indicate the sign of the charge generated at a particular position.

Table 3.1 shows the effect of solvent polarity on four different nucleophilic substitution reactions. Creation or destruction of charge gives the biggest effects; spreading or dispersal of charge as in the second and third examples in the table gives smaller effects. Molecular and radical reactions do not involve charge build-up in the transition state, and are little affected by solvents; thus a check for the presence or absence of a solvent effect often allows a distinction to be made between radical or molecular mechanisms, on the one hand, and polar reaction mechanisms, on the other. Care should be taken in applying a solvent effect test. The change in solvent should be small, such as a 10% increase in water in aqueous ethanol; more dramatic changes in solvent polarity risk changing the mechanism.

Table 3.1 Examples of solvent effects on reaction rates

Reagents		Transition state		Products	Type	Effect on rate of increasing solvent polarity
Me_3C-Br	$\rightarrow$	$Me_3C^+\ Br^-(+H_2O)$	$\rightarrow$	$Me_3C-OH_2^+ + Br^-$	Charge created	Large increase
$HO^- + CH_3-Br$	$\rightarrow$	$HO^{\delta-}\cdots CH_3\cdots Br^{\delta-}$	$\rightarrow$	$HO-CH_3 + Br^-$	Charge spread	Small decrease
$H_2O + {}^+SEt_3$	$\rightarrow$	$H_2O^{\delta+}\cdots Et\cdots^{\delta+}SEt_2$	$\rightarrow$	$H_2O^+-Et + SEt_2$	Charge spread	Small decrease
$HO^- + {}^+SEt_3$	$\rightarrow$	$HO^{\delta-}\cdots Et\cdots^{\delta+}SEt_2$	$\rightarrow$	$HO-Et + SEt_2$	Charge destroyed	Large decrease

3.3 Electronic Effects of Substituents in Polar Reactions

Solvent effects are useful in differentiating polar from non-polar reactions, but give no direct information about the location or polarity of the charges being produced or destroyed. The effect of substituent groups near to a reaction site provides more detailed insight.

A convenient system for studying substituent effects is the equilibrium between *meta-* and *para-*substituted benzoic acids and their corresponding anions (reaction 3.5). The acids are straightforward to synthesize, and the acidities in water at 25 °C are readily determined: the pH of a solution containing equal molar quantities of the acid and its fully ionized sodium salt will be equal to $-\log K_A$, where K_A is the dissociation constant of the acid. *Ortho* substituents are not considered because of complications caused by steric effects.

$$ (3.5) $$

$$ K_A = \frac{[H^+][XC_6H_4CO_2{}^-]}{[XC_6H_4CO_2H]} \qquad (3.6) $$

$$ \log_{10}\left(\frac{K_A}{K_H}\right) = \sigma \qquad (3.7) $$

We define a **substituent constant** σ (Greek sigma = s for *s*ubstituent) for any particular *meta* or *para* substituent as $\log(K_A/K_H)$, where K_A is the dissociation constant for the substituted benzoic acid and K_H is the dissociation constant for benzoic acid itself (equation 3.7). σ values for some common substituents are given in Table 3.2.

Table 3.2 Some values of Hammett substituent constants[a]

Substituent	meta σ	para σ	σ^+	σ^-
NMe$_2$	−0.10	−0.32	−1.70	–
Me	−0.06	−0.14	−0.31	–
OMe	0.10	−0.28	−0.78	–
F	0.34	0.15	−0.07	–
Cl	0.37	0.24	0.11	–
Br	0.37	0.26	0.15	–
I	0.34	0.28	0.13	–
CO$_2$R	0.35	0.44	–	0.74
CF$_3$	0.46	0.53	–	–
CN	0.62	0.70	–	0.88
NO$_2$	0.71	0.81	–	1.26

[a]For further values, see for example N. B. Chapman and J. Shorter, *Correlation Analysis in Chemistry*, Plenum, New York, 1978

Looking at reaction (3.5), the undissociated acid is uncharged, whereas the anion has a negative charge. We might therefore expect that electronegative substituents would withdraw electrons from the anion, therefore spreading the negative charge, stabilizing the anion, and thus making the compound more acidic. In accordance with this, the halogens all have positive σ values at the *meta* positions. However, there are anomalies: for example, σ_{para} for the very electronegative fluorine is only just positive, and the methoxy substituent appears to be electron withdrawing at the *meta* position but electron releasing if it is at the *para* position.

3.3.1 Inductive and Resonance Effects

To make sense of the σ values in Table 3.2, we need to look in more detail at how substituents can stabilize a developing positive or negative charge. Two factors are involved. The more obvious is the effect of the electronegativity of the substituent. Carbon and hydrogen have similar electronegativities, so C–H and C–C bonds in neutral saturated organic molecules are non-polar. Carbon becomes more electronegative in CH_3^+ as it changes its hybridization and acquires a positive charge. If one hydrogen is replaced by a fluorine atom, the electronegative fluorine atom will polarize the sigma bond so that, relative to the C–H bond, the electron density will be centred closer to the fluorine, and away from the C^+. This will build up the positive charge on the carbon and destabilize the ion. Conversely, groups such as alkyl groups can (relative to hydrogen) release electrons and stabilize a carbocation. This effect is called the **inductive effect**. Most of the common groups found in organic compounds are centred on elements which are more electronegative than carbon, and therefore tend to withdraw electrons inductively. The effect falls off sharply with distance.

The second and often more powerful influence is the **resonance** or **delocalization effect**. For any molecule for which more than one electron-pair bond "structure" can be written, the true structure will be intermediate between these "structures" and the molecule will be more stable than expected. Well-known examples include benzene (Figure 3.3a) and charged structures such as the allyl cation (Figure 3.3b). A carbocation can be stabilized either by an adjacent lone pair, which can be donated to form a double bond (Figure 3.3c) or by donation of electrons from an adjacent multiple bond (Figure 3.3b). Anions can be stabilized by donation of the lone pair to an adjacent multiply bonded atom (Figure 3.3d).

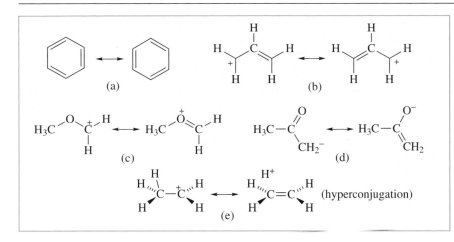

Figure 3.3   Delocalization

Resonance effects are frequently opposed to the inductive effect. The methoxy group is a striking example. The oxygen atom will withdraw electrons inductively but release them from its lone pair by resonance. The inductive effect predominates at the *meta* position, but the resonance effect predominates at the *para* position. Why is this?

The inductive effect falls off with distance, so we should expect that the inductive withdrawal of electrons by the methoxy group should be somewhat less at the *para* position (though still in the same direction). The resonance effect, on the other hand, will release electrons and produce negative charges at three locations on the benzene ring, as shown in Figure 3.4. For the *meta* substituents, none of these positions is adjacent to the carboxylate anion group, but for the *para* substituent, one of the negative charge positions is adjacent to the carboxylate anion as shown: this will destabilize the anion and outweigh the inductive withdrawal of electrons, giving a net negative σ value.

Inductive effects arise from the differing electronegativities of different atoms in the molecule. They operate through bonds or space and fall off rapidly with distance. Resonance effects, arising from interactions between non-bonding orbitals or π-bonds with a π-system, can be transmitted over long distances.

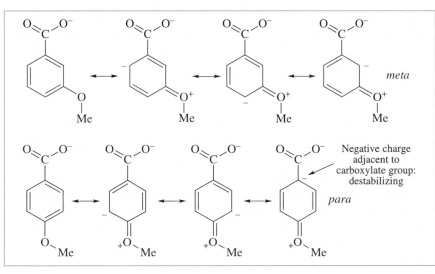

Figure 3.4   Resonance effects of *meta* and *para* substituents

The halogens also show inductive withdrawal of electrons and conjugative release. The release is greatest for fluorine in spite of its electronegativity. Being a first-row element, its size is similar to that of carbon, and overlap between its lone pair orbital and the neighbouring carbon p orbital is therefore greater than for chlorine. However, unlike the methoxy group, the resonance effect at the *para* position does not quite outweigh the inductive effect. The dimethylamino group shows the greatest difference between σ_{para} and σ_{meta}. Nitrogen, being more electropositive than oxygen, more readily accommodates a positive charge, so resonance from structures analogous to those shown in Figure 3.3(c) will be more effective. The CO_2Et, CN and NO_2 groups all have resonance effects in the same direction as the inductive effect; accordingly, the positive σ values at the *meta* positions are enhanced at the *para* position by the extra resonance contribution.

Finally, we consider the methyl substituent. The very slightly negative σ value at the *meta* position indicates that, relative to hydrogen, the methyl group releases electrons inductively to a neighbouring carbon atom. The more negative value at the *para* position indicates a conjugative release as well; this is surprising because the group is saturated and there is no lone pair to donate electrons. The effect is explained in terms of **hyperconjugation**. The alternative structure shown in Figure 3.3(e) is thought to contribute to the overall structure of a methyl group attached to a sp^2 hybridized carbon atom.

Thus from the experimentally obtained acidities of benzoic acids with a particular substituent at the *meta* and *para* positions, we obtain insight into the inductive and resonance effects for that substituent.

Worked Problem 3.1

Q Hammett σ constants for the NMe_3^+ substituent at the *meta* and *para* positions are 0.99 and 0.96, respectively. What can be inferred about the electronic effects of this substituent?

A σ_m reflects particularly the inductive effect of the substituent. The formal positive charge on the nitrogen makes this group strongly electron withdrawing, much more so than fluorine for example. The fact that σ_p is slightly smaller indicates that, unlike the NMe_2 group, NMe_3^+ has no lone pair to contribute electrons to the ring by the resonance effect, so the resonance contribution is virtually absent.

3.3.2 Hammett Reaction Constants: ρ (rho) Values

The Hammett σ substituent constants would be of little value if they only applied to the dissociation of substituted benzoic acids. Fortunately, this is not the case. For a very wide variety of reactions or equilibria involving a substituted phenyl group, if a plot is made of log k or log K against the Hammett σ substituent constant, a straight line is obtained, showing that a relationship of type (3.8) holds:

$$\log_{10}\left(\frac{k_A}{k_H}\right) = \rho\sigma \qquad (3.8)$$

This is the **Hammett equation**. The slope of the plot is denoted by ρ (Greek rho = r for reaction), and is characteristic of the reaction being studied. If negative charge is being built up in the transition state (or positive charge is being destroyed), ρ will be positive. If the reaction centre is further away from the benzene ring, as in the ionization of phenylacetic (phenylethanoic) acids (Table 3.3c), the ρ value will be smaller. If the reaction takes place even nearer the substituent on the benzene ring itself, as in electrophilic aromatic substitution, the value can be substantially higher than unity (Table 3.3i). Substituents have a greater effect in non-polar solvents or in the gas phase, where there is no solvent to stabilize developing charges (Table 3.3l).

Table 3.3 Hammett ρ values

Reaction	Medium	ρ
(a) $ArCO_2H \rightleftharpoons ArCO_2^- + H^+$	Water	1.00
(b) $ArCO_2H \rightleftharpoons ArCO_2^- + H^+$	Ethanol	1.96
(c) $ArCH_2CO_2H \rightleftharpoons ArCH_2CO_2^- + H^+$	Water	0.49
(d) $ArNH_3^+ \rightleftharpoons ArNH_2 + H^+$	Water	2.77
(e) $Ar_3CCl \rightleftharpoons Ar_3C^+ + H^+$	Liquid SO_2	−3.97
(f) $ArCMe_2Cl + H_2O \rightarrow ArCMe_2OH + HCl$	Acetone/water	−4.54
(g) $ArH + HNO_3 \rightarrow ArNO_2 + H_2O$	Acetic anhydride	−7.29
(h) $ArH + HOBr \rightarrow ArBr + H_2O$	$HClO_4$/dioxane/water	−6.2
(i) $ArH + Br_2 \rightarrow ArBr + HBr$	Acetic acid/water	−12.1
(j) $ArCl^a + MeO^- \rightarrow ArOMe + HCl$	Methanol	3.9
(k) $ArO^- + H_2C\overset{O}{\diagdown}CH_2 \rightarrow$ $ArO-CH_2-CH_2-O^-$	Ethanol	−1.12
(l) $ArCH_2^\bullet \rightarrow ArCH_2^+ + e^-$	Gas	−20
(m) $ArCH(Cl)CH_3 \rightarrow ArCHCH_2 + HCl$	Gas	−1.36
(n) $ArCH_3 + Cl^\bullet \rightarrow ArCH_2^\bullet + HCl$	CCl_4	−0.66
(o) $ArCOCl + Bu_3Sn^\bullet \rightarrow ArCO^\bullet + Bu_3SnCl$	m-Xylene	+2.6

[a] $ArCl = 1\text{-}Cl\text{-}2\text{-}NO_2\text{-}4\text{-}X\text{-}C_6H_3$

For reactions in which positive charge is being built up (or negative charge destroyed), the effect of substituents will be reversed, and substituents that release electrons will favour reaction. Thus in reactions of this sort, the slope of the Hammett plot will be negative, and negative values of ρ will result.

σ^+ and σ^-

Hammett plots should follow σ if the charge build-up or destruction is at a site not conjugated with the substituent. If conjugated, σ^+ is followed by *para* substituents which can stabilize a positive charge, σ^- by *para* substituents which can stabilize a negative charge.

The defining reaction for Hammett σ constants is the ionization of substituted benzoic acids. Although a full negative charge is produced in the anion, which is delocalized between the two oxygen atoms of the carboxylate anion group, the charge cannot be spread by delocalization onto the benzene ring or further onto substituents (reaction 3.9a). Substituents such as CH_3CO stabilize the anion by an electron-withdrawing resonance contribution which builds up a partial positive charge on the carbon atom adjacent to the CO_2^-, but there is no through conjugation of the type that is possible for the substituted phenoxide ion. For this ion, structure **1**, in which the negative charge is moved to the carbonyl group with no net loss of covalent bonds or charge creation, will contribute more than structure **2** does to the stabilization of the *para*-acetylbenzoate ion, which would involve charge creation and net breakage of a covalent bond.

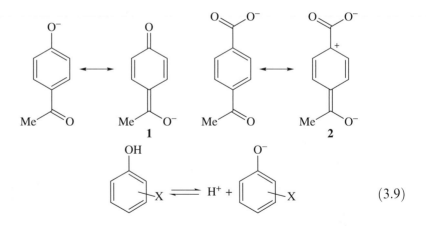

$$(3.9)$$

This means that substituents that can withdraw electrons conjugatively will have a greater effect on the acidity of phenols (reaction 3.9) than on the acidity of benzoic acids. The difference will only be appreciable at the *para* position; for the *meta* position, no through resonance is possible. If we make a plot of $\log K_A$ for the acidities of *meta*-substituted phenols against Hammett σ constants, we find that the points lie on a straight line, the slope of which gives the ρ value for the reaction as $+2.23$

(see Figure 3.5a). For *para* substituents such as p-NO_2, which stabilize the anion significantly by through conjugation, the acidities are much greater than expected on the basis of their σ values. For each substituent, a new σ^- value can be defined, to ensure that the point will fall on the graph. These σ^- values will apply for reactions in which a product anion can be stabilized by through-bond conjugation.

In an analogous manner, through conjugation from *para* substituents can stabilize a positive ion. The *para*-methoxy substituent shows this type of stabilization for the cumyl (2-phenylprop-2-yl) cation **3**. Rates of hydrolysis of substituted cumyl chlorides (reaction 3.10) in aqueous acetone were plotted against σ. The *meta*-substituted points fell on the line; substituents such as methoxy and methyl were faster than expected, and these values were used to derive σ^+ values for *para* substituents of this type (Figure 3.5b).

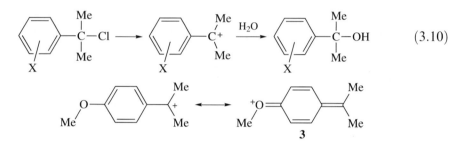

$$(3.10)$$

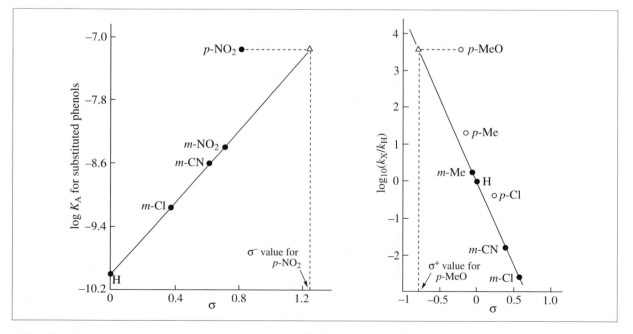

Figure 3.5   (a) Hammett plot for acidity of phenols. (b) Hammett plot for reaction (3.10)

3.3.3 Deductions about Mechanism from Substituent Studies

Ideally, we determine the Hammett ρ factor for a reaction by studying the rates of reaction with a number of *meta* substituents, plotting log k against Hammett σ substituent constants. A positive ρ factor means that negative charge is being built up (or positive charge is being destroyed). The magnitude of ρ often provides information about how close to the benzene ring is the position where charge is being built up.

We now look at the rates for a number of *para* substituents, choosing ones which have different σ, σ^+ and σ^- values. We check how well the experimental log k values for these *para*-substituted compounds lie on the line defined by the *meta* substituents when we use (a) σ, (b) σ^+ or (c) σ^- values for the *para* substituents. If σ^- is followed, this provides strong evidence that a negative charge is being built up at a position in the molecule which is conjugated with the *para* substituent, if ρ is positive. If ρ is negative, a negatively charged site of this type must be being destroyed during the reaction. If σ^+, is followed, with a negative ρ value, a positive charge is being built up during the reaction at a position conjugated with the *para* substituent; if ρ is positive, this positive charge, present in the reactant, must be destroyed in the reaction. If σ is followed, the positive or negative charge being built up or destroyed during the reaction is not conjugated with the reaction centre. The various possibilities are listed in Table 3.4, with examples.

Table 3.4   Deductions made from the sign of ρ, and which of σ, σ^+ or σ^- gives the best fit in the Hammett equation[a]

	ρ positive	ρ negative
σ^-	Reaction centre[b] conjugated with Ar group; negative charge builds up in the transition state. Example: reaction (3.9)	Reagent stabilized by conjugative electron release of electrons to Ar group; this conjugation is less important in the transition state. Example: reaction (k) in Table 3.3
σ	Reaction centre not conjugated with Ar group; transition state has less demand for electrons than reactant. Example: hydrolysis of aromatic esters by bases	Reaction centre not conjugated with Ar group; transition state has a greater demand than reactant for electrons. Example: hydrolysis of arylmethyl acetates, $ArCH_2OCOMe$, by acids
σ^+	Reaction centre conjugated with Ar group and stabilized by electron-releasing substituents; conjugation less important in transition state. Examples: rare	Reaction centre conjugated with Ar group; transition state has greater demand than reactant for electrons. Example: reaction (i) in Table 3.3

[a]σ values are used where a special σ^+ or σ^- value has not been derived
[b]The reaction centre for the purpose of this table is the position at which charge is built up or destroyed during the reaction

The following worked problems show how useful information about reaction mechanisms is derived from the sign and magnitude of the Hammett ρ factor, and whether σ, σ^+ or σ^- is followed for *para* substituents.

Worked Problem 3.2

Q The ionization constants for $ArCH_2CH_2CO_2H$ in water show a Hammett ρ value of 0.24 and follow σ rather than σ^+ or σ^-. Comment on these observations in relation to other data in Table 3.3.

A Entries (a) and (c) in Table 3.3 show Hammett ρ values for $ArCO_2H$ and $ArCH_2CO_2H$ of 1.0 and 0.49, respectively, showing that the insulating effect of the CH_2 group reduces ρ by a factor of about 2. $ArCH_2CH_2CO_2H$, with a further CH_2 group, shows a further fall by a factor of 2, showing a similar insulating effect of the second CH_2 group. The σ dependence is expected for substituents that are not conjugated with the negative charge of the carboxylate anion.

Worked Problem 3.3

Q The bromination of substituted benzenes in acetic acid/water mixtures shows a ρ value of −12.1 and a dependence on σ^+ rather than σ or σ^-. What conclusions can be drawn?

A The very large negative ρ value indicates a large build up of positive charge close to the substituent in the transition state. The dependence on σ^+ shows that the position at which the charge is being formed is conjugated with substituents at the *para* position. This supports (but does not prove) the formation of a charged intermediate of the type shown below.

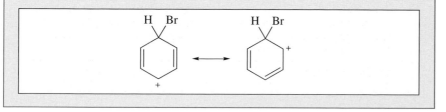

3.4   Steric Effects

The previous section has focused on the electronic effects of substituents. The size of substituents can also affect reaction rates. Bulky substituents destabilize molecules or ions, and the greater the number of groups attached to a reaction centre, the greater the destabilization.

Many reactions involve a change in coordination round the reaction centre. If the coordination number increases on going from the starting molecule to the transition state, replacement of an atom or group at the reaction centre by a bulkier group will cause a reduction in rate. If there is a reduction in coordination number, introduction of a bulkier group will cause an increase in rate. These two effects are termed **steric retardation** and **steric acceleration**, respectively.

A good example of steric retardation is shown by S_N2 reactions, in which an incoming nucleophile displaces a leaving group (reactions 1.1a and 3.11). In the transition state, the new bond is partially formed whilst the breaking bond is still partially bonded. Thus the original four coordination round the carbon reaction centre is increased in the transition state to five coordination. Hence if we increase the bulk of the substituents, the crowding will be worse in the transition state, so the reaction will be slower. For the reaction of sodium ethoxide with primary bromoalkanes $(R^1R^2R^3C)CH_2Br$ in ethanol at 55 °C (reaction 3.11), the effects on the relative rates of reaction for increasing numbers of substituent methyl groups are given in Table 3.5.

five-coordinate
transition state

(3.11)

Table 3.5   Relative effects of successive substitutions

	Reaction (3.11) (S_N2)	Reaction (3.12) (S_N1)	Acid strength of substituted acetic acids
Substituent change	H → Me	Me → t-Bu	H → Cl
Effect of first substitution	÷3.6	×1.2	×80
Effect of second substitution	÷9.4	×15.2	×24
Effect of third substitution	÷7022	–	×6
Increasing or decreasing?	Increasing	Increasing	Decreasing
Steric or electronic?	Steric	Steric	Electronic

The effect of introducing a methyl group at the β-position is to decrease the rate by a factor of 4. The second methyl produces a more pronounced reduction by a factor of 9 as the transition state becomes more crowded. The third methyl group produces a massive rate decrease by a factor of 7000; this last methyl effectively blocks access to the ethoxide ion as it approaches the reaction centre.

In a S_N1 reaction (1.1b or 3.12), the reaction centre in the halogenoalkane is four coordinated. The transition state is close to the intermediate carbocation in structure, which is three coordinated (and planar). Crowding is therefore less severe in the transition state than in the initial molecule, so substitution of bulky groups at the reaction centre gives rise to steric acceleration. In reaction (3.12), the solvolysis of R^1R^2MeCCl in 50% ethanol at 25 °C, relative rates for Me_3CCl, t-$BuMe_2CCl$ and t-Bu_2MeCCl are 1:1.21:18.4. There is a very small increase in rate as the first t-butyl group is introduced, but a much larger effect is caused by the second t-butyl group. This is seriously crowded by the first t-butyl group in the original molecule, but the crowding is relieved as the bond angles at the central carbon atom spread out from about $109\frac{1}{2}°$ to $120°$.

$$R^1{}_{\cdots}\overset{\text{Me}}{\underset{R^2}{\diagup}}Cl \quad \xrightarrow[\substack{\text{EtOH} \\ H_2O}]{S_N1} \quad \overset{\text{Me}}{\underset{R^1\;\;R^2}{\diagup}}{}^{+} \quad \longrightarrow \quad \begin{array}{c} R^1R^2(\text{Me})COH \\ + \\ R^1R^2(\text{Me})COEt \end{array} \qquad (3.12)$$

This increase in effect on rates as further similar substituents are introduced is typical of steric effects, where each further substituent causes an increase in the amount of crowding. This is to be contrasted with polar effects, where a second substituent usually causes a smaller relative increase or decrease than the first. For example, if the substituent stabilizes a positive charge by releasing electrons, the charge is diminished and a further substituent will be releasing electrons to a centre where there is less demand, so the stabilizing effect will be less. In the series acetic (ethanoic) acid, chloroacetic acid, dichloroacetic acid and trichloroacetic acid, the acid dissociation constant values (K_A) in water are 1.76×10^{-5}, 1.40×10^{-3}, 3.32×10^{-2} and 2×10^{-1}, respectively. The first chlorine strengthens the acid by a factor of 80, the second only produces an increase of a factor of 24, and the third produces less than a factor of 10. This suggests that the effect of the chlorine substitution is electronic rather than steric. These steric and electronic substituent effects are summarized in Table 3.5.

When substituents are added successively at or near to a reaction centre, if the effect of adding successive identical substituents causes increasing changes in the rate, the effect is probably steric. If the effect diminishes, the origin of the effect is probably electronic.

Worked Problem 3.4

Q The Friedel-Crafts acylation of alkylated benzenes by $[PhCO^+][SbF_6^-]$ (reaction 3.13) shows the following percentage distributions of isomers. As the number of methyl substituents on the α-carbon atom is increased, the proportion of *ortho* isomer decreases. Is this likely to be an electronic or a steric effect?

$$C_6H_5R + PhCO^+SbF_6^- \longrightarrow PhCOC_6H_4R + HSbF_6 \qquad (3.13)$$

$R = CH_3, CH_2Me, CHMe_2, CMe_3:$

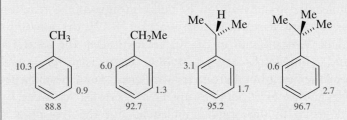

A As we introduce the first extra methyl group into the toluene molecule, the *ortho* percentage drops by a factor of 1.7($=10.3/6.0$). The second group produces a decrease of a factor of 1.9, but the third produces a much larger decrease, by a factor of 5.2. The greater effect of the second and especially the third substituent methyl group supports a steric explanation, consistent with the three-coordinate *ortho* carbon atom becoming four coordinated in the transition state as the new group starts to bond.

3.5   Stereochemistry

The stereochemistry of products relative to starting materials can give important clues to the structure of the transition state. The stereochemistry of molecules gives rise to two types of isomerism: **optical isomerism (enantiomerism)** and **geometrical isomerism**.

3.5.1 Enantiomerism (Optical Isomerism)

If four different groups surround a central carbon atom, two (and only two) different stereoisomers are possible, as illustrated in Figure 3.6. These are known as enantiomers (or optical isomers). These two isomers are different because rotation about any of the bond axes of the molecule cannot bring all four groups of the *R*-isomer into coincidence with the *S*-isomer. There are not more than two: any other arrangement of groups round the central carbon atom can always be rotated around a bond axis to bring the ethyl group to the top, and if the hydrogen atom is not at the back, rotation round the C–ethyl bond to bring the hydrogen to the back will lead to one or the other of the two isomers shown in Figure 3.6.

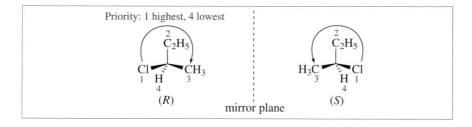

Priority: 1 highest, 4 lowest

mirror plane

Figure 3.6 Enantiomers

To enable the enantiomers to be identified without the need for a drawing, the following convention has been established. The four groups are assigned a priority from 1 (highest) to 4 (lowest) on the basis of the atomic numbers of the atom directly connected to the central carbon atom. This establishes the chlorine atom as the group with highest priority (1) and hydrogen as the lowest (4), but both methyl and ethyl both have a carbon atom attached to the central carbon atom, with the same atomic number. Ties are broken by considering the atoms attached to the first (carbon) atom in the group. Ethyl has a carbon and two hydrogens (C,H,H); methyl has three hydrogens (H,H,H). Since carbon has a higher atomic number than hydrogen, the ethyl group has the higher priority (2) and methyl the lower (3). A single atom of higher priority takes precedence over any number of atoms of lower priority, *e.g.* (N,H,H) beats (C,C,C). If the tie is not broken at the atoms two removed from the central atom, we go one further out and so on until the tie is resolved. A double bond is counted as having a normal bond to the atom at the other end, plus a further bond to a fictitious atom of the same type which does not have any further connections. Thus isopropyl [–CH(CH$_3$)$_2$] (C,C,H) and vinyl [–CH=CH$_2$] (C,(C),H) both beat ethyl [–CH$_2$CH$_3$] (C,H,H). Isopropyl [–CH(CH_3)$_2$] beats vinyl [–CH=CH_2] by having six hydrogens at the next position out, whereas vinyl has only two. These slightly arcane rules are to ensure compatibility with the older D,L stereochemical convention used for sugars and α-amino-acids.

Once priorities have been assigned, the molecule is viewed from a position where the group of lowest priority (4) is at the back. The remaining groups are viewed in decreasing order of priority: (1) to (2) to (3). If this is clockwise, the configuration is (R) (*rectus*, right-handed). If anticlockwise, the configuration is (S) (*sinister*, left-handed).

Enantiomers have very similar properties. Since each has the same groups attached to the same atoms, with identical bond lengths and angles, properties such as melting point, solubility, reaction rate with particular reagents, *etc.*, will be identical. The two enantiomers are related as an object and its mirror image as shown in Figure 3.6, and can only be distinguished under circumstances in which this "handedness" or **chirality** is important. For present purposes, the most important difference is that solutions of the enantiomers will rotate the plane of polarized light in opposite directions by an exactly equal amount. The rotation of a 1 g mL^{-1} solution in a 10 cm cell, measured for yellow sodium D-line light, is called the specific rotation and is denoted as $[\alpha]_D^{20}$. The superscript number refers to the temperature used in °C. Positive values for $[\alpha]_D^{20}$ correspond to a clockwise rotation, negative values to anti-clockwise rotation.

For mechanistic studies, we are mostly interested in relative configuration. We would like to know, for example in an S_N2 reaction, whether the incoming group replaces the outgoing group without changing stereochemistry at the reaction centre (**retention**), or if the reaction involves **inversion** at the reaction centre to give the opposite configuration. Unfortunately, the sign of rotation is not a reliable guide to absolute or relative configuration. Of 19 common optically active α-amino acids, $RCH(NH_2)CO_2H$, involved in proteins which all have the same configuration (Figure 3.7), 11 have negative specific rotations but 8 have positive rotations.

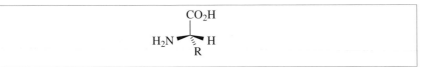

Figure 3.7 Configuration of naturally occurring α-amino acids

Absolute configurations can now be obtained by X-ray crystallography. Determination of the absolute stereochemistry of both reagent and product molecules will in principle allow us to determine if a particular reaction involves retention or inversion. Historically, an alternative, which does not require knowledge of absolute stereochemistry, involves carrying out a set of reactions in which an optically active reagent molecule is converted in several steps via a number of intermediates back into the same molecule, but with the opposite configuration. All but one of the reaction steps do not take place at the optically active centre, and in these steps it is assumed that no change in configuration can take place. This leaves one step only where reaction takes place at the optically active centre. If the final molecule has the same rotation as the initial molecule, the reaction must take place with retention of configuration; if the sign of rotation is reversed, the reaction must involve inversion of configuration to the other enantiomer.

This type of logic is illustrated by reaction sequence (3.14), in which the positively rotating form of the alcohol **4** is converted in two stages through **5** and **6** into its negatively rotating enantiomer **7**. At the time these experiments were carried out, no absolute stereochemistries were known, but since the overall process gave inversion of configuration, either one or all three of the individual steps must involve inversion of configuration. Step (a) involves attack of the hydroxyl oxygen on the *p*-toluenesulfonyl chloride, and should not affect the C–O bond, and step (c) involves ester hydrolysis, which again is thought to leave the C–O bond intact. This leaves step (b), in which an acetate ion attacks the chiral carbon atom, displacing the *p*-toluenesulfonate anion, and so this reaction is where the inversion of stereochemistry must occur.

$$
\begin{array}{ccccc}
\text{Me} & & \text{Me} & & \text{Me}\\
| & \xrightarrow[\text{TsCl}]{a} & | & \xrightarrow[\text{$^-$OAc}]{b} & |\\
\text{PhCH}_2-\text{CH}-\text{OH} & & \text{PhCH}_2-\text{CH}-\text{OTs} & & \text{PhCH}_2-\text{CH}-\text{OAc}\\
\mathbf{4}\ \alpha = 33.0° & & \mathbf{5}\ \alpha = 31.1° & & \mathbf{6}\ \alpha = -7.1°
\end{array}
$$

$$
c\ \Big|\ {}^-\text{OH} \qquad (3.14)
$$

$$
\begin{array}{c}
\text{Me}\\
|\\
\text{PhCH}_2-\text{CH}-\text{OH}\\
\mathbf{7}\ \alpha = -32.2°
\end{array}
$$

Ts = p − MeC$_6$H$_4$SO$_2$−;
Ac = MeCO − .

A simple way of determining the stereochemical course of S$_N$2 reactions is to study reactions of disubstituted cyclopentane and cyclohexane rings, where retention or inversion can be established by finding out if the product has the same geometric form (*cis* or *trans*, see next section) as the starting molecule, which proves retention of configuration at the reaction centre, or has the opposite geometrical form, which proves inversion. Reaction (3.15) shows that inversion of configuration takes place for this S$_N$2 reaction.

$$(3.15)$$

Transition State for S$_N$2 Reactions

The simplest way to account for the inversion of configuration which always accompanies S$_N$2 reactions is that the incoming nucleophile approaches the molecule from the side opposite to the leaving group (Figure 3.8). As the nucleophile approaches, the bond to the leaving group lengthens and the hybridization of the central carbon atom moves towards sp^2, so that the unchanged groups take up a trigonal configuration with 120° bond angles. In the transition state **9**, the remaining p orbital allows a partial bond to both the incoming and the leaving group, at a 180° angle to each other. As the reaction goes beyond the transition state, the hybridization reverts to sp^3, but the bond to the incoming nucleophile is on the opposite side of the molecule to where the leaving group was situated. The movement of the three unchanged groups round the central carbon atom is like an umbrella blowing inside out in the wind, and is called inversion. The product **10** has the opposite configuration to that of the starting molecule **8**. This mechanistic insight into the S_N2 reaction can *only* be obtained from stereochemical studies.

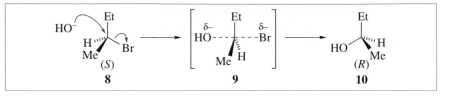

Figure 3.8 Inversion in S_N2 reactions

Stereochemistry of S_N1 Reactions

In contrast to S_N2 reactions, S_N1 reactions of molecules with a chiral centre proceed with racemization, that is the product comprises a mixture of equal numbers of molecules with the same and opposite configurations compared with the starting molecule. Each individual reaction must of course involve either retention or inversion of configuration, but the processes which lead to these two products must proceed at the same rate. This accords with the mechanism proposed in Chapter 1. The rate-determining step involves the heterolytic breakage of the bond to the leaving group, leaving a carbocation which from this and other evidence has a planar trigonal structure with sp^2 hybridization. Since the carbocation is planar, the incoming nucleophile can approach from either face equally readily, as shown in Figure 3.9, giving rise to an equal mixture of (R) and (S) molecules, whose rotations will cancel out. A mixture of this type is known as a **racemic mixture**, and racemization is the characteristic stereochemical feature of S_N1 reactions.

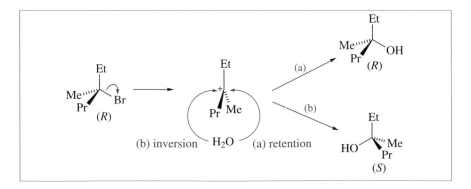

Figure 3.9 Stereochemistry of S_N1 reactions

In contrast to S_N2 reactions, where the stereochemical outcome is always complete inversion, the racemization of S_N1 reactions is sometimes only partial. This can arise because the leaving group, though free from the carbocation intermediate, still blocks that side of the molecule, so that attack from the opposite face is easier, giving a product in which the inverted molecules predominate over those with retention of stereochemistry.

Neighbouring Group Assistance

A vast number of S_N2 reactions at carbon centres has been studied. In every case, it is found that inversion of configuration occurs. This certainty now allows us to use inversion of configuration as a diagnostic test: if inversion does not take place, the reaction cannot have a simple S_N2 mechanism. S_N1 reactions show partial or complete racemization, depending on how "free" the intermediate carbocation becomes.

A small number of nucleophilic substitutions show retention of configuration, *i.e.* the new group comes in on the same side as the leaving group departs, in spite of the unfavourable steric interactions that would occur between the two groups. Reactions of this type tend to be faster than expected by comparison with similar compounds. The reactions are kinetically of first order, depending only on concentration of the halogen compound, and not on that of the attacking nucleophile. These reactions are characterized by the presence of a neighbouring substituent group (*e.g.* Br, OH) which has a lone pair that makes the group nucleophilic. This group can carry out an intramolecular substitution, as shown in reaction (3.16), forming a cyclic intermediate **11**, with the expulsion of the leaving group and inversion of configuration at the reacting centre. The incoming nucleophile attacks the intermediate by an S_N2 reaction, opening the ring, restoring the internal substituent group and inverting the configuration again at the reaction centre. The overall reaction involves two inversions of configuration, thereby giving overall retention of configuration.

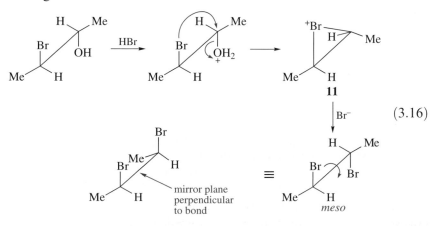

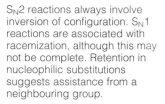

(3.16)

S_N2 reactions always involve inversion of configuration. S_N1 reactions are associated with racemization, although this may not be complete. Retention in nucleophilic substitutions suggests assistance from a neighbouring group.

In these "sawhorse" representations of stereochemistry, the part of the molecule on the lower left is in front of the part on the upper right.

3.5.2 Geometrical Isomerism

A second form of stereoisomerism is associated with restriction of rotation round a bond joining two atoms in a molecule. The restriction can be caused by a double bond, or by a ring structure, and gives rise to two **geometrical isomers**. For a 1,2-disubstituted ethene, the substituents can be on the same side of the double bond (*cis*) or on opposite sides (*trans*). The two geometrical isomers will have different reactivity and physical properties. For example, the two geometrical isomers of 1,2-dichloroethene (Figure 3.10a and b) differ by 11 °C in boiling points, attributable to the presence of a dipole moment in the *cis* isomer, whereas the *trans* isomer has no overall dipole. The *cis*/*trans* nomenclature can be extended to tri- or tetrasubstituted ethenes, provided that there is a pair of identical substituents on each end of the double bond, for example **13** in equation (3.18) below. For more complex situations, where the double bond has three or four different substituents, the isomers are distinguished by assigning priorities as for the *R*,*S* convention to the substituents at each end of the double bond. If the substituents of higher priority at each end of the double bond are on the same side, the molecule is *Z* (*zusammen*, German, together); if they are on opposite sides, the isomer is *E* (*entgegen*, German, opposite) (Figure 3.10c and d). For double bonds with identical substituents on each end, *Z* ≡ *cis* and *E* ≡ *trans*.

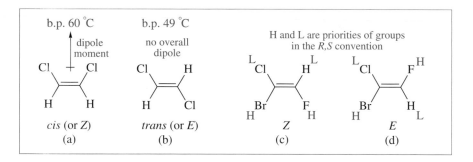

Figure 3.10 Geometrical isomerism

Geometrical isomerism is a useful tool in investigating the stereochemical course of addition and elimination reactions.

Stereochemistry of E2 Elimination Reactions

For S_N2 reactions, there is almost always a competing reaction in which the nucleophile removes a proton from the β-carbon atom. This is accompanied by loss of the leaving group and formation of a C=C double bond, as shown in reaction (3.17). The reaction is usually elementary, and is termed **E2** or **bimolecular elimination**. In an E2 reaction, there is no intermediate. Can we determine anything about the transition state, in particular, the direction of approach of the incoming nucleophile (base) from the stereochemical course of the reaction?

$$\text{(3.17)}$$

Compounds such as PhCHBr–CHBrPh have two asymmetric centres, each of which can have the R or the S configuration. There will be three stereoisomers (reactions 3.18 and 3.19), which can be designated S,S (**14**), R,R (**15**), and R,S (**12**) (S,R will be the same as R,S because of the symmetry of the molecule). The R,R and S,S forms will be optically active and rotate polarized light in opposite directions by an equal amount. Synthesis will normally give a 50/50 mixture of the two isomers, the **racemic** or ($\pm$) mixture. The R,S or *meso* isomer will be optically inactive since it is identical with its mirror image (the mirror plane in **12** can be seen in the eclipsed conformation obtained by rotation of the front of the molecule clockwise by 60° round the central C–C bond).

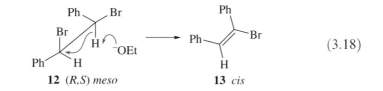

12 (*R,S*) *meso*   **13** *cis*

$$\text{(3.18)}$$

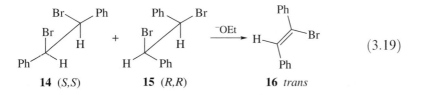

14 (*S,S*)   **15** (*R,R*)   **16** *trans*

$$\text{(3.19)}$$

Experimentally, it is found that PhCHBr–CHBrPh reacts with sodium ethoxide by elimination of hydrogen bromide to give 1-bromo-1,2-diphenylethene. The *meso* compound **12** gives the *cis*-alkene **13** (reaction 3.18), whereas the (S,S and R,R) racemic mixture **14** and **15** gives the *trans*-alkene **16** (reaction 3.19). The disposition of the substituent groups in the alkene product suggests that the molecule reacts in the conformation shown in Figure 3.11 and the incoming base attacks *anti* to the departing halide anion (180° dihedral angle). In this conformation the carbon atoms which are rehybridizing from sp^3 to sp^2 can begin to form the π bond as the reaction proceeds, as shown in Figure 3.11. Since two new bonds are being formed at the same time as two are being broken, the reaction is concerted and can take place with a relatively low activation energy.

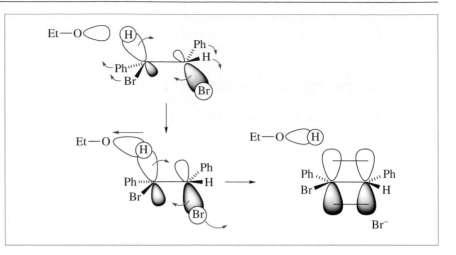

Figure 3.11 *Anti* elimination in
E2 reactions

The geometrical isomerism of the
products of elimination reactions
shows that, in E2 reactions,
reaction is fastest if the leaving
groups are *anti* to each other. If an
*anti*arrangement is impossible,
syn elimination takes place.

For rigid strained molecules such as the norbornane derivative **17**, it is impossible for the leaving proton and toluenesulfonate anion to adopt the *anti* conformation, since the molecule is locked in an eclipsed conformation where the only possible dihedral angles are 0° (*syn*) or 120°. Under these circumstances, *syn* elimination in reaction (3.20) allows the reaction to take place in a concerted manner, and is the preferred route. It must be emphasized that for compounds in which free rotation is possible, *anti* elimination takes place almost exclusively, even though both *syn* and *anti* orientations would allow concerted reaction. The *anti* orientation has the advantage of having the attacking base on the opposite side of the molecule to the leaving halide, and the staggered conformation involved would be an energy minimum, compared with the energy maximum for the eclipsed conformation required for *syn* elimination.

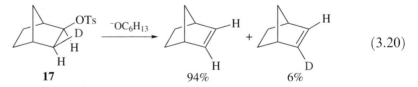

$$\tag{3.20}$$

Summary of Key Points

1. Solvent effects provide information about the relative polarity of the transition state compared with the reactants, and may be used to cut down mechanistic possibilities.

2. The electronic effects of substituents (electron-withdrawing or -releasing) give more precise information: Hammett ρ values show whether positive or negative charge is being built up (or destroyed) at the reaction centre, and the dependence of $\log_{10} k$ or $\log_{10} K$ on σ, σ^+ or σ^- tells us whether or not this charge build up is in a position conjugated with the substituent.

3. The two types of electronic effects of substituents are inductive and resonance; these can operate in the same or opposite directions.

4. The steric effect of introducing substituents at the reaction centre is to increase the rate if the transition state is less crowded than the reactant, and to decrease the rate if the transition state is more crowded.

5. Optical and geometrical isomerism throw light on the stereochemistry of reactions. S_N2 reactions always involve inversion of configuration; S_N1 reactions are associated with racemization, though this may not be complete; retention in nucleophilic substitutions suggests assistance from a neighbouring group.

6. In E2 reactions, reaction is fastest if the leaving groups are *anti* to each other. If an *anti* arrangement is impossible, *syn* elimination takes place.

Further Reading

C. Reichardt, *Solvent Effects in Organic Chemistry*, Verlag Chemie, Weinheim, 1979.

C. D. Johnson, *The Hammett Equation*, Cambridge University Press, Cambridge, 1973.

E. L. Eliel, S. H. Wilen and L. N. Mander *Stereochemistry of Carbon Compounds*, Wiley, New York, 1994 (long).

M. J. T. Robinson, *Organic Stereochemistry*, Oxford University Press, Oxford, 2000 (short).

O. B. Ramsay, *Stereochemistry*, Heyden, London, 1981 (historic).

Problems

3.1. For the bimolecular hydrolysis of 2-bromopropane with OH^- in aqueous ethanol, the rate constant decreased by a factor of 1.6 when the proportion of water was raised from 20% to 40%. When a corresponding change in solvent was made for the bimolecular hydrolysis of 2-bromopropane with water, the rate increased by a factor of 2.8. Explain these differences.

3.2. Hammett σ_m values for the OMe and O–CO–Me groups are 0.10 and 0.26, respectively. Comment on this difference.

3.3. Suggest a number of substituents for which (a) σ^+ will be significantly different from σ and (b) σ^- will be significantly different from σ.

3.4. Aldehydes XC_6H_4CHO undergo addition of hydrogen cyanide across the C=O double bond in the presence of aqueous sodium cyanide, Na^+CN^-. Two possible mechanisms are shown below. For each of the two routes, predict the sign of the Hammett ρ factor and whether the Hammett plot will follow σ, σ^+ or σ^-.

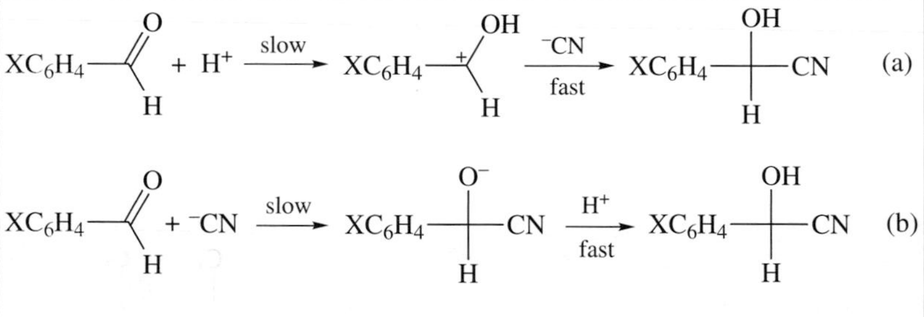

3.5. 1-Arylethyl acetates, $MeCO_2CH(Me)C_6H_4X$, decompose on heating in the gas phase in a first-order reaction to give acetic acid and the arylethene. For substituents on the aromatic ring, a Hammett correlation with σ^+ is observed, with $\rho = -0.7$. What light does this throw on the mechanism of the decomposition?

3.6. Suggest reactions other than those discussed in this chapter which should show (a) positive and (b) negative ρ values and for which correlations would be best (1) with σ, (2) with σ^+ or (3) with σ^-, i.e. six categories in all. It may be difficult to find examples of some of these categories.

3.7. For the S_N1 solvolysis of the tertiary chloroalkanes R–Cl in 50% ethanol at 25 °C, relative rates for three different halides are 1:1.2:18.4 for R = Me$_3$C, t-BuMe$_2$C, and t-Bu$_2$MeC, respectively. Are these differences likely to be due to electronic or to steric effects?

3.8. The acid dissociation constants of CH_3CO_2H, $PhCH_2CO_2H$, and Ph_2CHCO_2H in water are 1.76×10^{-5}, 5.2×10^{-5} and 1.15×10^{-4}, respectively. Is the effect of the phenyl group primarily steric or primarily electronic?

3.9. Using the rules to establish the priority of substituents, confirm that the compounds in Figure 3.8 have been assigned their correct stereochemical configurations.

3.10. All 19 common optically active α-amino acids involved in proteins have the same configuration shown in Figure 3.7, and were given the designation L in the D,L convention, applicable to sugars and α-amino acids. Typical side chains are R = PhCH$_2$ for phenylalanine and HO$_2$CCH$_2$CH$_2$ for glutamic acid. All but one of the α-amino acids have the (S) configuration. Cysteine, with R = HSCH$_2$, has the same configuration shown in Figure 3.6, but is designated (R). Why is this?

3.11. Three of the stereoisomers of 1,2,3,4,5,6-hexachlorocyclo-hexane, **X**, **Y** and **Z**, undergo elimination of HCl in 76% aqueous ethanol with second-order rate constants of 2.1×10^{-4}, 0.15 and $0.5 \ M^{-1} \ s^{-1}$, respectively. Suggest a reason for the differences in rates.

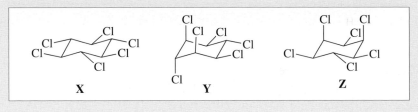

X   Y   Z

4
Anions and Nucleophilic Reactions

Chapter 3 considered elementary reactions and the information about their transition states that can be obtained by a variety of techniques. The next three chapters deal with complex reactions, in which intermediates are produced. These may be short lived, but they have an independent existence, show spectroscopic properties and can sometimes be trapped out of the reaction mixture.

This chapter considers reactions in which anionic intermediates are involved. **Anions** are species that carry a negative charge. Also important here are:

Bases: species (often anions) that can remove a proton from another molecule.

Nucleophiles: species with a lone pair of electrons that can be donated to an electron-deficient centre, normally a carbon atom, to form a covalent bond.

Aims

After reading this chapter you should be able to:

- Define and distinguish between bases and nucleophiles
- Assess the likely points of attack by nucleophiles and bases on molecules
- Suggest plausible reaction steps for condensation reactions
- Understand the catalytic role of bases in base-catalysed reactions
- Rationalize the steps involved in substitution reactions of carboxylic acid derivatives
- Predict whether a particular carbanion will undergo elimination to give a carbene

4.1   Acids and Bases

Bases are molecules or ions that abstract protons. **Acids** can donate protons to a base. The base formed when an acid loses a proton is called the **conjugate base** of the acid and *vice versa* (reaction 4.1). The strength of an acid corresponds to its ability to donate protons; that of a base depends on its ability to accept protons. It follows that if an acid is strong (readily donates protons to a different base), its conjugate base is weak (cannot readily abstract protons from the conjugate acid of the different base). A strong acid will react with a strong base to give a weak acid and a weak base, as shown in reaction (4.2), with an example shown in reaction (4.3) where hydrogen chloride, which is a stronger acid than phenol, donates a proton to the sodium salt of phenol to give phenol and the chloride ion. In practical terms, this reaction goes entirely to completion.

$$\underset{\text{acid}}{\text{H—A}} \;\rightleftharpoons\; \text{H}^+ \;+\; \underset{\substack{\text{conjugate} \\ \text{base}}}{\text{A}^-} \tag{4.1}$$

$$\underset{\substack{\text{strong} \\ \text{acid}}}{\text{H—A}} \;+\; \underset{\substack{\text{strong} \\ \text{base}}}{\text{A}'^-} \;\rightleftharpoons\; \underset{\substack{\text{weak} \\ \text{acid}}}{\text{H—A}'} \;+\; \underset{\substack{\text{weak} \\ \text{base}}}{\text{A}^-} \tag{4.2}$$

$$\text{H—Cl} \;+\; \underset{\substack{\text{(sodium)} \\ \text{phenoxide}}}{\text{}^-\text{O—Ph}} \;\rightleftharpoons\; \underset{\text{phenol}}{\text{H—O—Ph}} \;+\; \text{Cl}^- \tag{4.3}$$

   Table 4.1 shows the acidities of a number of acids, arranged in order from strongest to weakest. A number of factors contribute to acid strength. These include the electronegativity of the atom which bears the negative charge in the anion formed on loss of the proton, the strength of the bond to the proton, and delocalization. Other things being equal, electronegative atoms, in particular the halogens and oxygen, will be more stable carrying a negative charge, and these elements comprise the central atom of the conjugate bases of the strongest acids. Electronegativity will not account for the strengths of the hydrogen halides: hydrogen fluoride is the weakest of these acids, even though fluorine is the most electronegative element. However, the hydrogen–fluorine bond is one of the strongest known, whereas the hydrogen–iodine bond is extremely weak, and this factor counterbalances the electronegativity effect.

Table 4.1 Acidity, basicity and nucleophilicity

Acid (strongest at top, weakest at bottom)	Conjugate base (weakest at top, strongest at bottom)	pK_A of acid[a] (most negative = most acidic)	Nucleophilicity, n, of base[b] (high = most nucleophilic) (bold = oxygen-centred nucleophile)
$HClO_4$	ClO_4^-	-11	
HI	I^-	-10	5
H_2SO_4	HSO_4^-	-10	
HBr	Br^-	-9	3.5
HCl	Cl^-	-7	2.7
H_3O^+	H_2O	-1.7	**0**
HF	F^-	3.2	2.0
$MeCO_2H$	$MeCO_2^-$	4.8	**2.7**
H_2S	HS^-	7.0	5.1
PhOH	PhO^-	9.9	**3.5**
$MeCOCH_2COMe$	$MeCOCH^-COMe$	9	
HCN	CN^-	9.2	5.1
RNH_3^+	RNH_2	10–11	4.5
$MeCOCH_2CO_2R$	$MeCOCH^-CO_2R$	11	
$EtOCOCH_2CO_2Et$	$EtOCOCH^-CO_2Et$	13	
$HCCl_3$	CCl_3^-	13.6	
H_2O	OH^-	15.7	**4.2**
EtOH	EtO^-	16	
cyclo-C_5H_6	cyclo-$C_5H_5^-$	16	
$RCOCH_2R$	$RCOCHR^-$	19–20	
$ROCOCH_2R$	$ROCOCHR^-$	24.5	
$H-C{\equiv}C-H$	$H-C{\equiv}C^-$	25	
$PhNH_2$	$PhNH^-$	30.6	
NH_3	NH_2^-	38	
$PhCH_3$	$PhCH_2^-$	40	
$CH_2{=}CH\text{-}CH_3$	$CH_2{=}CH\text{-}CH_2^-$	43	
C_6H_6	$C_6H_5^-$	43	
$CH_2{=}CH_2$	$CH_2{=}CH^-$	44	
CH_4	CH_3^-	48	

[a] $pK_A = -\log_{10}K_A = -\log_{10}\{[H^+][A^-]/[HA]\}$. For strong, highly dissociated, acids, the equilibrium value of $[H^+]$ is high, giving high K_A values, and thus low or negative pK_A values
[b] Nucleophilicity, n, is defined by the expression $\log_{10}(k_n/k_0) = sn$, where (k_n/k_0) is the relative rate of attack by the nucleophile on a substrate compared with attack by water on the same substrate, s is the sensitivity of the substrate to nucleophilic attack, and n_{H_2O} is defined as zero

For acids where a hydrogen–oxygen bond is to be broken, we might expect similar acidities, and indeed this is true for compounds such as water and ethanol where there are no structural stabilizing features in either the acid or the corresponding conjugate base. However, for acetic (ethanoic) acid, the acetate ion has two equally contributing resonance structures (see Chapter 3). The negative charge will therefore be spread

from the carbonyl group. This is the basis of the Michael addition reaction (4.8). The ethoxide ion abstracts a proton from diethyl malonate (propane-1,3-dioate) to give the malonate anion **1** which can attack the butenone molecule to give the stabilized anion **2**, which in turn abstracts a proton from an ethanol molecule to give the adduct molecule **3**.

$$CH_2(CO_2Et)_2 \; \overset{\text{}}{\underset{\text{}}{\rightleftharpoons}} \; \overset{\text{}^-OEt/HOEt}{} \; ^-CH(CO_2Et)_2$$
$$\mathbf{1}$$

$$Me-\overset{\overset{O}{\|}}{C}-CH{=}CH_2 \;\; ^-CH(CO_2Et)_2 \longrightarrow Me-\overset{\overset{O^-}{|}}{C}{=}CH-CH_2-CH(CO_2Et)_2 \qquad (4.8)$$
$$\mathbf{1} \qquad\qquad\qquad\qquad \mathbf{2}$$

$$Me-\overset{\overset{O}{\|}}{C}-CH_2-CH_2-CH(CO_2Et)_2 \; \overset{\text{HOEt}}{\underset{\text{}}{\rightleftharpoons}} \; Me-\overset{\overset{O}{\|}}{C}-\overset{-}{C}H-CH_2-CH(CO_2Et)_2$$
$$\mathbf{3} \;\; 70\%$$

(d) Fragmentation of anions.

Elimination of a stable molecule, often CO_2 or a carbonyl compound, can give an anion which usually reacts further. Examples include elimination of carbon dioxide from acetates (reaction 4.9). The reaction of iodine with acetone (propanone) in the presence of base also involves a step of this type (reaction 4.17), as does the reverse of the Claisen condensation (reaction 4.34).

$$CH_3-\text{\Large\diagup}\hspace{-0.6em}\overset{\overset{O}{\diagdown}}{\underset{O}{}} \longrightarrow CO_2 \; + \; ^-CH_3 \; \overset{H_2O}{\longrightarrow} \; CH_4 \; + \; ^-OH \qquad (4.9)$$

4.3.2 Structure of Carbanions

The methide anion (CH_3^-) has been shown to be pyramidal in the gas phase, and other saturated carbanions are also pyramidal. Hybridization is essentially sp^3, with the lone pair electrons occupying one of the tetrahedral positions.

Conjugated carbanions, for example those with an adjacent carbonyl group, are essentially planar. Hybridization is sp^2 with π overlap to the carbonyl group, and the majority of the negative charge resides on the oxygen atom. ^{13}C NMR spectra of anions show upfield shifts compared with the corresponding hydrocarbons, consistent with the increased shielding caused by the electron pair on the carbon atom which bears the charge. The ^{13}C NMR spectrum of conjugated anions such as penta-2,4-dienide (Figure 4.1) shows that the charge is mainly spread on the odd-numbered carbon atoms, in accordance with resonance or molecular orbital theory, with the even-numbered atoms having shifts in the range

Carbanions have a formal negative charge on a carbon atom. They are strongly basic and nucleophilic.

shown by alkenes, showing that little of the negative charge is associated with these atoms.

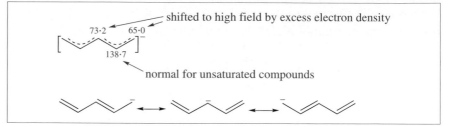

Figure 4.1 Chemical shifts in the ^{13}C NMR spectrum of the penta-2,4-dienide anion

4.3.3 Reactions of Carbanions (and Nucleophiles)

Carbanions, being strong bases and nucleophiles, react readily with acids and undergo the S_N2 reactions considered in Chapter 1 and the bimolecular E2 eliminations considered in Chapter 3. Other important reactions of carbanions (and nucleophiles) include addition to unsaturated compounds, especially those containing carbonyl groups, fragmentation (the reverse of addition) and elimination to give carbenes.

Reaction with Acids

This has been discussed earlier in the chapter. Carbanions will abstract a proton from acids (including carbon acids) provided that the acid is stronger than the conjugate acid of the carbanion (see Table 4.1). Since water is acidic, most organometallic compounds will react with water to give the hydrocarbon and the hydroxide ion, so water must be rigorously excluded from most organometallic preparations and reactions. The reaction of water with organometallic compounds does however provide a route to organic compounds deuterated (or tritiated) in particular positions, for example reaction (4.10).

$$D{-}\overset{\delta-}{\underset{}{O}}{-}\overset{\delta+}{\underset{}{D}}\quad\overset{\delta-}{\underset{}{R}}{-}\overset{\delta+}{\underset{}{Mg}}{-}Br \longrightarrow DO^- + D{-}R + {}^+Mg{-}Br \qquad (4.10)$$

Although the reaction of hydroxide or alkoxide ions with carbonyl compounds to abstract a proton from a carbon atom adjacent to the carbonyl group is thermodynamically unfavourable, and the equilibria in reactions such as (4.11) and (4.12) lie well over to the left, the high reactivity of the carbanion intermediates formed makes reactions of this type of vital importance in a number of cases, especially in halogenation (reaction 4.14) and in some carbonyl addition reactions, *e.g.* reactions (4.20) and (4.30).

$$H{-}O^- + H{-}CH_2CHO \rightleftharpoons H{-}O{-}H + {}^-CH_2CHO \qquad (4.11)$$

$$Et{-}O^- + H{-}CH_2CO_2Et \rightleftharpoons Et{-}O{-}H + {}^-CH_2CO_2Et \qquad (4.12)$$

Nucleophilic Substitution

Carbanions are strong nucleophiles. The S_N2 reaction with halogeno-alkanes gives a new carbon–carbon bond, allowing long or complex hydrocarbon groups to be introduced into organic molecules (reaction 4.13). Carbanions will react with halogens to give the halogenoalkane and halide ion (4.14).

$$Na^+ R^{\frown} R'\!-\!\overset{\frown}{Hal} \longrightarrow Na^+ Hal^- + R\!-\!R' \qquad (4.13)$$

$$R\!-\!CO\!-\!CH_2^{\frown} Hal\!-\!\overset{\frown}{Hal} \longrightarrow R\!-\!CO\!-\!CH_2\!-\!Hal + Hal^- \qquad (4.14)$$

Worked Problem 4.2

Q   When acetone is allowed to react with iodine in the presence of sodium hydroxide, the final major products are triiodomethane and the acetate anion. How do you account for this?

A   The first step is clearly the formation of iodoacetone (reaction 4.15, a specific example of reaction 4.14). In iodoacetone, there are still five replaceable hydrogen atoms, so the base can attack to give either CH_3COCHI^- or $^-CH_2COCH_2I$. Since iodine is more electronegative than hydrogen, the former anion will be more stabilized, because the negative charge is partly located on the carbon with the iodine and thus can be stabilized by inductive electron withdrawal. Therefore further iodination will preferentially give 1,1,1-triiodoacetone (reaction 4.16). The hydroxide ion adds to the C=O carbon atom to give the intermediate anion **4**, which eliminates a $^-CI_3$ anion, a strong base which will abstract a proton from a water molecule to give CHI_3 (reaction 4.17). The elimination of $^-CI_3$ is possible because the electronegative iodine atoms inductively withdraw negative charge from the carbon, thus stabilizing the carbanion.

$$CH_3COCH_3 \underset{H_2O}{\overset{^-OH}{\rightleftharpoons}} CH_3COCH_2^- \overset{I_2}{\longrightarrow} CH_3COCH_2I + I^- \qquad (4.15)$$

$$CH_3COCH_2I \longrightarrow \longrightarrow CH_3COCI_3 \qquad (4.16)$$

$$(4.17)$$

E2 Eliminations

Alkyl sodium and potassium compounds, R^-Na^+ and R^-K^+, are strong enough bases to react with ethers, such as diethyl ether (ethoxyethane), to give ethene and the ethoxide ion (reaction 4.18). This means that ethers are unsuitable solvents for the preparation of these reactive compounds.

$$R^- \quad H-CH_2-CH_2-OEt \longrightarrow R-H + CH_2{=}CH_2 + {}^-OEt \qquad (4.18)$$

Worked Problem 4.3

Q Although alkylsodium compounds react with diethyl ether, organomagnesium halides (Grignard reagents) do not. Why is this?

A Alkylsodium compounds are almost completely ionic and the alkyl anions are very strong bases, capable of abstracting a proton from ethers to cause an E2 elimination, as in reaction (4.18). Alkylmagnesium halides have polarized C–Mg bonds; although there is considerable ionic character in the bond, it is less ionic than in the alkylsodium compounds. Accordingly, the organomagnesium halides are less basic than alkylsodium compounds, and although they will react with water to give the alkane (reaction 4.10), they are not basic enough to remove a proton from diethyl ether.

Addition to the Carbonyl Group

The carbonyl group is strongly polarized in the direction ${}^{\delta+}C{=}O^{\delta-}$, so it is not surprising that a carbanion (or any other nucleophile) will bring its electron pair to the ${}^{\delta+}C$ atom to form a bond, displacing the π bond electron pair onto the oxygen atom which can readily accommodate them, as shown in reaction (4.19). This type of reaction is important synthetically and biologically. An important example is the aldol condensation, the simplest example of which is shown in reaction (4.20). The hydroxide ion reversibly removes the weakly acidic proton from an acetaldehyde (ethanal) molecule to give the stabilized carbanion **5**. Although present only in a tiny concentration at equilibrium, this carbanion is highly nucleophilic and adds to the carbonyl carbon atom of a second acetaldehyde molecule to give an adduct that reversibly reacts with water to give aldol (3-hydroxybutanal, **6**). The mechanism shown in reaction (4.20) is supported by the observed kinetics: first order with respect to [OH$^-$], second order with respect to [CH$_3$CHO]. A rapid equilibrium is set up between the acetaldehye and its anion **5** (equation 4.20a), giving a small concentration of $^-CH_2CHO$, which is proportional to [CH$_3$CHO][OH$^-$]. The product is formed by a second-order reaction of the anion **5** with a further acetaldehyde molecule (equation 4.20b), thereby accounting for the overall kinetics.

$$R \overset{}{\underset{}{\big\backslash}} C \overset{}{=} O \longrightarrow \overset{R}{\underset{O^-}{C}} \qquad (4.19)$$

$$HO^- + CH_3-CHO \overset{water}{\rightleftharpoons} H_2O + {}^-CH_2-CH{=}O \longleftrightarrow CH_2{=}CH-O^- \qquad (4.20a)$$
$$\underset{5}{}$$

$$\overset{-CH_2-CH=O}{\underset{Me \rightleftharpoons CH=O}{}} \longrightarrow \underset{H \quad O^-}{\overset{Me}{\underset{}{}}\underset{}{C}} \overset{CH_2-CH=O}{\underset{}{}} \overset{H_2O}{\rightleftharpoons} \underset{H \quad OH}{\overset{Me}{\underset{6}{}}\underset{}{C}} \overset{CH_2-CH=O}{\underset{}{}} + {}^-OH \qquad (4.20b)$$

Neutral nucleophiles, such as Grignard reagents (R–Mg–Br), also react readily with carbonyl compounds, as shown in reaction (4.21). The hydroxide ion will also react as a nucleophile (reaction 4.22), but here the overall equilibrium lies well to the left so that only a few percent of the diol **7** is present in an aqueous solution of acetone.

$$\underset{Me_2C=O}{Et{-}Mg{-}Br} \longrightarrow \underset{Me \quad O^- \, {}^+Mg{-}Br}{\overset{Me \quad Et}{C}} \overset{H^+}{\underset{H_2O}{\longrightarrow}} \underset{Me \quad OH}{\overset{Me \quad Et}{C}} \qquad (4.21)$$

$$\underset{Me}{\overset{Me \quad {}^-OH}{C{=}O}} \rightleftharpoons \underset{Me \quad O^- \, H}{\overset{Me \quad OH}{C}} \rightleftharpoons \underset{Me \quad OH}{\overset{Me \quad OH}{\underset{7}{C}}} \qquad (4.22)$$

Reaction with Esters

Nucleophiles react with esters by substitution rather than addition, as shown in ester hydrolysis (4.23) or conversion to an amide by ammonia (4.24). Why the difference from aldehydes and ketones, and what is the mechanism?

$$MeCO_2Et + {}^-OH \longrightarrow MeCO_2{}^- + EtOH \qquad (4.23)$$

$$MeCO_2Et + NH_3 \longrightarrow MeCONH_2 + EtOH \qquad (4.24)$$

The key difference is thermodynamic. By placing a carbonyl group and a hydroxyl group on the same carbon atom, we have created a new functional group, the carboxylic acid group. This has different chemical properties and, in particular, is more thermodynamically stable than would be expected from a compound containing separate C=O and

Carboxylic acids and their derivatives (*e.g.* esters and amides) possess thermodynamic stabilization. They tend to react by substitution to retain this stabilization, rather than by addition, which would destroy it.

OH groups. The magnitude of this stabilization can be estimated by considering the hypothetical reaction (4.25) in which the hydroxyl group in ethanol and the aldehyde hydrogen in ethanal are notionally exchanged to give acetic acid and ethane. If there were no interaction between the C=O and the OH group in acetic acid, the reaction would be expected to be thermoneutral. In fact, based on experimental heats of formation, the reaction is 115.5 kJ mol^{-1} exothermic. This shows that there is a strong stabilizing interaction between the C=O and the OH groups. A lone pair on the oxygen atom of the OH group can interact with the π bond of the C=O group, or, to use resonance language, there are two possible resonance contributions to the structure. Structure **9**, which has the same number of covalent bonds as **8**, but has charge separation, will not contribute as much as **8** to the overall structure, but the infrared carbonyl stretching frequency ranges of aldehydes and ketones, on the one hand, and carboxylic acids and derivatives, on the other, demonstrate less double-bond character in carboxylic acids and derivatives, showing a significant contribution of **9** to the structure. Stabilization of this type is also important for other related functional groups, in particular esters ($-CO_2R$) and amides ($-CONH_2$).

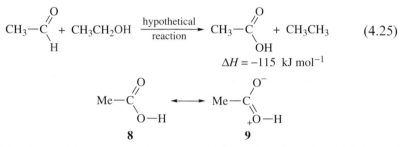

$$\Delta H = -115 \text{ kJ mol}^{-1}$$

Aldehydes and ketones tend to react with nucleophiles by addition of the nucleophile at the carbonyl carbon, and this is often followed by addition of an electrophile at the carbonyl oxygen, giving a stable adduct. Reaction (4.26), involving an aldehyde and the cyanide ion, is a good example. The first step of reaction of the hydroxide ion with ethyl acetate (reaction 4.27a) likewise involves addition of the hydroxide ion to the ester carbonyl group to give the tetrahedral intermediate **10**. This destroys the carboxylic stabilization energy which would not be recovered if this intermediate were to add a proton. Instead, the lone pair electrons on the negatively charged oxygen come in to re-form the carbonyl double bond, and the ethoxide ion is expelled. This restores the carboxylic stabilization energy. In subsequent steps, the ethoxide ion reacts with a water molecule to give ethanol and the hydroxide ion (reaction 4.27b), which in turn removes a proton from the acetic acid to give the even more stabilized acetate anion, where the negative charge is delocalized equally onto its two oxygen atoms (reaction 4.27c).

$$\underset{\text{H}}{\overset{\text{Me}}{>}}\text{C}=\text{O} \quad \overset{\cap}{\text{CN}} \longrightarrow \underset{\text{H}}{\overset{\text{Me}}{>}}\text{C}\underset{\text{CN}}{\overset{\text{O}^-}{<}} \quad \overset{\text{H}^+}{\longrightarrow} \quad \underset{\text{H}}{\overset{\text{Me}}{>}}\text{C}\underset{\text{CN}}{\overset{\text{OH}}{<}} \qquad (4.26)$$

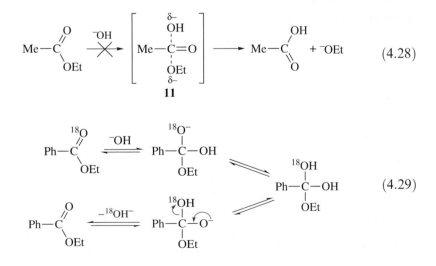

$$\text{Me}-\overset{\text{OH}}{\underset{\text{OEt}}{\overset{|}{\underset{|}{\text{C}}}}}\;\;\overset{\text{O}}{\cdots} \;\;\rightleftharpoons\;\; \text{Me}-\overset{\text{OH}}{\underset{\text{OEt}}{\overset{|}{\underset{|}{\text{C}}}}}-\text{O}^- \;\;\longrightarrow\;\; \text{Me}-\overset{\text{OH}}{\underset{\text{O}}{\overset{|}{\underset{\|}{\text{C}}}}} \;+\; {}^-\text{OEt} \qquad (4.27\text{a})$$

$$\textbf{10}$$

$$\text{H}_2\text{O} + {}^-\text{OEt} \;\rightleftharpoons\; \text{HO}^- + \text{HOEt} \qquad (4.27\text{b})$$

$$\text{Me}-\overset{\text{OH}}{\underset{\text{O}}{\overset{|}{\underset{\|}{\text{C}}}}} + \text{OH}^- \quad\overset{\text{essentially}}{\underset{\text{irreversible}}{\longrightarrow}}\quad \text{Me}-\overset{\text{O}^-}{\underset{\text{O}}{\overset{|}{\underset{\|}{\text{C}}}}} \;\longleftrightarrow\; \text{Me}-\overset{\text{O}}{\underset{\text{O}^-}{\overset{\|}{\underset{|}{\text{C}}}}} + \text{H}_2\text{O} \qquad (4.27\text{c})$$

$$\text{stabilized}$$

Why not postulate the direct substitution reaction analogous to the S_N2 reaction, with a transition state **11** as shown in reaction (4.28), with no intermediate? The best evidence for the formation of a tetrahedral intermediate comes from studies of hydrolysis, in ordinary water ($H_2{}^{16}O$), of benzoate esters which have been labelled at the carbonyl oxygen with ^{18}O. If the reaction was not carried out to completion, and the unchanged ester was isolated, it was found that some of the ^{18}O had been lost from the carbonyl group. Quantitative analysis for ethyl benzoate showed that the ratio of the rate constant for exchange of oxygen to that for hydrolysis, k_{ex}/k_{hyd}, was 4.8, implying that an adduct is formed initially, but after a movement of a proton, the addition is reversed with loss of $^{18}OH^-$, as in reaction (4.29).

$$\text{Me}-\overset{\text{O}}{\underset{\text{OEt}}{\overset{\|}{\underset{\backslash}{\text{C}}}}} \;\overset{{}^-\text{OH}}{\cancel{\longrightarrow}}\; \left[\text{Me}-\overset{\overset{\delta-}{\text{OH}}}{\underset{\underset{\delta-}{\text{OEt}}}{\overset{\vdots}{\underset{\vdots}{\text{C}}}}}=\text{O}\right] \;\longrightarrow\; \text{Me}-\overset{\text{OH}}{\underset{\text{O}}{\overset{|}{\underset{\|}{\text{C}}}}} + {}^-\text{OEt} \qquad (4.28)$$

$$\textbf{11}$$

$$\underset{\text{OEt}}{\overset{{}^{18}\text{O}}{\text{Ph}-\text{C}}} \;\overset{{}^-\text{OH}}{\rightleftharpoons}\; \text{Ph}-\overset{{}^{18}\text{O}^-}{\underset{\text{OEt}}{\overset{|}{\underset{|}{\text{C}}}}}-\text{OH} \;\rightleftharpoons\; \text{Ph}-\overset{{}^{18}\text{OH}}{\underset{\text{OEt}}{\overset{|}{\underset{|}{\text{C}}}}}-\text{OH}$$

$$\underset{\text{OEt}}{\overset{\text{O}}{\text{Ph}-\text{C}}} \;\overset{-{}^{18}\text{OH}^-}{\rightleftharpoons}\; \text{Ph}-\overset{{}^{18}\text{OH}}{\underset{\text{OEt}}{\overset{|}{\underset{|}{\text{C}}}}}-\text{O}^- \qquad (4.29)$$

The Claisen Condensation

An important reaction of esters is the **Claisen condensation**, illustrated in reaction (4.30), in which two molecules of an ester react, in the presence of base, to give a β-keto ester. The initial steps of the reaction are similar to the aldol condensation already considered, but because of the lesser acidity of the ester compared with an aldehyde or ketone, a strong base has to be used, and the reaction cannot be carried out in water.

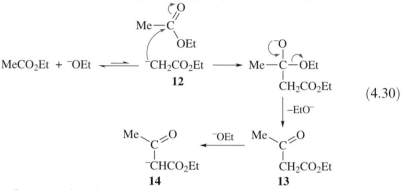

$$(4.30)$$

The first step involves the abstraction of a proton by the base from the α-carbon atom of the ester. The equilibrium for this reaction lies overwhelmingly in favour of starting materials, but the carbanion **12** produced is very reactive, and adds to the carbonyl group of an ester molecule. The tetrahedral intermediate loses an ethoxide ion to give the β-keto ester product **13**. This molecule is much more acidic than the original ester, and readily loses a proton to form the highly stabilized anion **14**. In practical terms, the reaction is carried out by adding sodium to the neat ester. A trace of alcohol is always present in the ester, which reacts with sodium to form the base sodium ethoxide. The ethanol produced in the reaction reacts with sodium as it is formed to provide further ethoxide ions, driving the reaction to the right, giving as the final product the sodium salt of the β-keto ester **14**, from which the β-keto ester is recovered at the end of the reaction by acidification.

Although this reaction cannot be carried out in aqueous solution in the laboratory, an enzymic version of the Claisen condensation is responsible in living cells for the conversion of acetic acid derivatives to fatty acids.

Base-catalysed Reactions

A number of reactions of carbonyl compounds require the presence of a base, but the base is not used up in the reaction and is therefore acting as a catalyst. The aldol condensation considered earlier (reaction 4.20) is an example, and another is the base-catalysed exchange of ^{18}O between water and esters (reaction 4.29) which accompanies hydrolysis. The base reacts with the carbonyl compound in the first stage of the reaction, either by abstraction of a proton (4.20) or by addition at the carbonyl carbon atom (4.29). In both cases, the base is regenerated in the final step of the reaction. Other reactions which similarly involve a base include ester hydrolysis (reaction 4.27) and the Claisen condensation (reaction 4.30). However, in these reactions, a mole of base is used up in the final step to convert the carboxylic acid to its anion in ester hydrolysis and the β-keto ester to its anion in the Claisen condensation. Since base is used up, these reactions are better described as **base-promoted**.

Reaction with Alkenes

Although carbanions and nucleophiles add readily to the highly polarized C=O group, addition to the C=C bond is difficult because of the high electron density in the π bond, which makes approach of nucleophiles difficult. We have already seen in reaction (4.8) that attack is possible if electron density is pulled away from one end of the C=C double bond by a carbonyl group. Organometallic compounds with a high degree of ionic character in the carbon–metal bond, such as butyllithium, will react with conjugated alkenes such as buta-1,3-diene or styrene (phenylethene) to give adducts which in the absence of moisture or oxygen can add further alkene molecules to give polymers; these continue to grow until all the alkene is used up. At this point, the **living polymer** is still reactive: if a further supply of the same or a different conjugated alkene is added, polymerization continues, giving in the latter case a **block copolymer** where blocks of different monomer units can be assembled (reaction 4.31). The process is finished by adding water or air, which quenches the organometallic compound by donation of a proton to give the alkane (reaction 4.32), or addition to a dioxygen molecule to give in turn the peroxide anion and then the alkoxide anion (reaction 4.33).

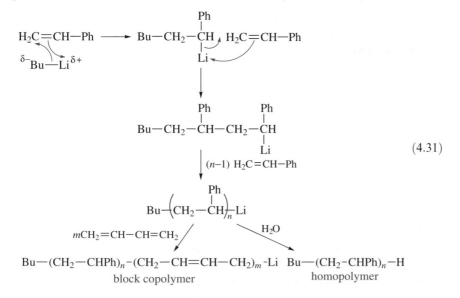

(4.31)

$$R\!-\!Li \xrightarrow{\ H_2O\ } R\!-\!H + Li^+ + {}^-OH \qquad (4.32)$$

$$R\!-\!Li \xrightarrow{\ O=O\ } R\!-\!O\!-\!O^-\,Li^+ \xrightarrow{\ R\!-\!Li\ } 2\,R\!-\!O^- + 2\,Li^+ \qquad (4.33)$$

Fragmentation

Carbanions (and other anions) can undergo fragmentation reactions to produce a carbanion and a smaller molecule, usually carbon dioxide or a molecule containing a carbonyl group. Examples include the loss of carbon dioxide from the acetate anion (reaction 4.9), considered above. Other important examples include reactions of type (4.34), which are the reverse of addition reactions such as (4.20) and (4.30).

$$(4.34)$$

Elimination to give Carbenes

Carbanions with a good leaving group, such as a halogen atom, attached to the carbanion centre may react in a unimolecular elimination process to give a **carbene**, a different type of reactive intermediate with only six electrons in the valence shell. Dichlorocarbene may be conveniently made by the reaction of potassium *t*-butoxide with trichloromethane (chloroform; reaction 4.35). Abstraction of the proton is made easier by the presence of the three chlorine atoms which stabilize the charge on the intermediate carbanion **15**, and the subsequent elimination of the chloride ion transfers the negative charge from carbon to the much more electronegative chlorine.

$$t\text{-BuO} + HCCl_3 \longrightarrow t\text{-BuOH} + {}^-\overset{Cl}{\underset{Cl}{C}}\!-\!Cl \longrightarrow :\overset{Cl}{\underset{Cl}{C}} + Cl^- \qquad (4.35)$$

$$\mathbf{15}$$

Carbenes are neutral reactive species that can form two bonds, either by insertion into a single covalent bond, or by addition to both ends of a double bond to give a cyclopropane ring.

Carbenes are very reactive: since they contain divalent carbon, they have the capacity to react to form two new bonds. Two important ways in which this happens are by addition to both ends of a double bond synchronously to give a cyclopropane ring (reaction 4.36) or by insertion into a single bond (reaction 4.37). The former reaction has synthetic utility, since few reactions exist to produce cyclopropane rings in good yields.

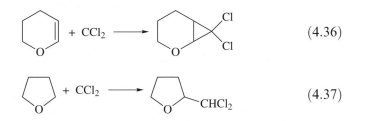

$$(4.36)$$

$$(4.37)$$

4.3.4 Detection of Carbanions and Indications for Anionic Intermediates

In some cases, carbanion intermediates can be detected, for example by NMR spectroscopy. More frequently, they are inferred as being the most likely intermediate, or characterized by trapping with reagents known to react with anions. Reactions taking place in basic solutions are likely to involve anionic intermediates, formed either by abstraction of a proton or by addition to a multiple bond (usually containing a heteroatom). Reactions in acidic media virtually never involve reactive anionic intermediates; any anions present merely balance the positive charges. Stepwise reactions in basic media usually involve alternating sequences of neutral and negatively charged species. Do not suggest steps involving positively charged species including protons: any protons needed to complete a reaction will come from solvent molecules.

For example, both cyclohexa-1,3-diene and its isomer cyclohexa-1,4-diene, when heated separately with t-$C_5H_{11}OK$ and t-$C_5H_{11}OD$ at 95 °C [t-C_5H_{11} = $(C_2H_5)(CH_3)_2C$–] give the same mixture of monodeuteriated compounds **16** and **17** (reaction 4.38). This is not the equilibrium mixture, which suggests that the mixture of products arises from an intermediate which is the same for both reactions. The only likely intermediate is the stabilized anion **18**, arising from proton abstraction from the cyclohexadiene molecules by the base, followed by abstraction of a deuteron from the solvent by the common intermediate anion.

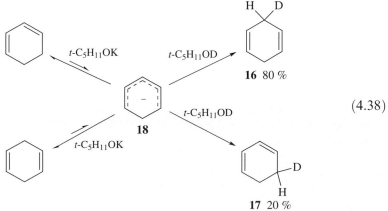

$$(4.38)$$

Trapping of anions can be achieved by adding methyl methacrylate (methyl 2-methylpropenoate, **19**) to a reaction mixture. The formation of a polymer, precipitated out of the reaction mixture by adding methanol, indicates an anion intermediate (reaction 4.39). Addition of a mixture of methyl methacrylate and styrene is even more specific for reactive anions. If an anion initiates the polymerization, the polymer contains only methyl methacrylate units. If a radical is involved, a copolymer of styrene and methyl methacrylate will be formed, whereas cations will give a styrene polymer. The three possible polymer outcomes, and thus the nature of the initiator, can readily be distinguished by combustion analysis: the weight percentages of carbon and hydrogen in the three types of polymers are markedly different (see Table 4.2).

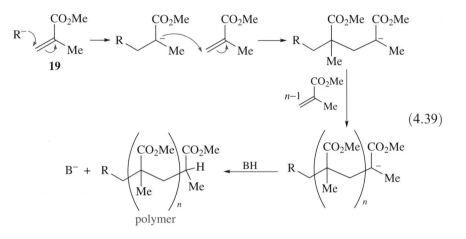

(4.39)

Table 4.2 Copolymerization of methyl methacrylate in the presence of various initiators

Polymer composition	Carbon (%)	Hydrogen (%)	Initiator
Polystyrene	92.3	7.7	Cation
Poly(methyl methacrylate)	60.0	8.1	Anion
Copolymer of styrene and methyl methacrylate (1:1)	76.4	7.9	Radical

Worked Problem 4.4

Q Pyridine-2-carboxylic acid (**20**) decarboxylates on heating, to give pyridine and carbon dioxide. The reaction is more rapid for **20** than for its isomers **21** and **22**. Suggest a mechanism for this reaction and an experiment to support it.

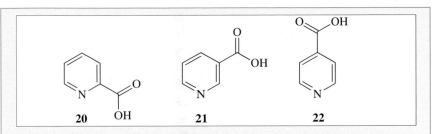

A   The fact that **20** reacts faster than **21** or **22** suggests that the first step may be the abstraction of a proton from the carboxylic acid group by the nitrogen atom to give the zwitterion **23** (reaction 4.40). This could be followed by loss of carbon dioxide from the carboxylate group to give **24**, which would rapidly be converted into pyridine by a proton shift.

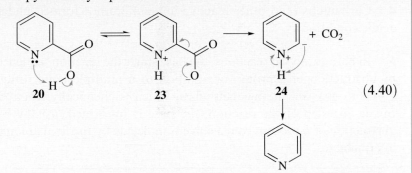

$$(4.40)$$

The presence of **24** as an intermediate is supported by carrying the reaction out in the presence of acetophenone (phenylethanone), when the adduct **25** is isolated (reaction 4.41). A carbanion intermediate such as **24** would be expected to add to the carbon atom of a carbonyl group, giving the adduct **25**.

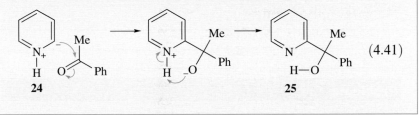

$$(4.41)$$

Summary of Key Points

1.   Bases are species that can donate a lone pair of electrons to a proton, forming a covalent bond, to give its conjugate acid. If the conjugate acid is a weak acid, the base will be a strong base. Bases are usually negatively charged or neutral.

2. Nucleophiles can donate a lone pair of electrons to form a bond to an atom (usually carbon), by displacing a leaving group (nucleophilic substitution), by breaking a multiple bond (nucleophilic addition) or by adding to a carbocation (second step of an S_N1 reaction).

3. Carbanions can exist as metal salts, and are also important intermediates in reactions carried out under basic conditions. They react as strong bases and nucleophiles, undergoing hydrogen abstraction reactions, nucleophilic substitution and addition, fragmentation, and loss of halide ions to give carbenes. They can be identified spectroscopically or by trapping.

4. Carbanions (and other reactive anions) are not formed under acidic conditions.

5. In base-catalysed reactions, the base starts the reaction sequence by abstraction of a proton or addition to a multiple bond (often C=O) to give an intermediate anion which is chemically reactive. In the last step of the reaction, the base is re-formed, usually by abstraction of a proton from a solvent molecule by the final anionic intermediate.

Further Reading

M. B. Smith and J. March, *March's Advanced Organic Chemistry*, Wiley, New York, 2001, Chapter 5.

Problems

4.1. Define the terms base, nucleophile, carbanion and carbene. List the common reactions of carbanions and carbenes.

4.2. In Worked Problem 4.1 it was asserted that the Ph_3C^- anion is more stable than the Ph_2CH^- anion because of the extra stabilization caused by delocalization of the charge over the extra benzene ring. Draw and count the resonance structures for the two anions, with the charge on different carbon atoms, to confirm this.

4.3. The pK_a values for CH_3CH_3, $CH_3CO_2C_2H_5$ and CH_3COCH_3 are 48, 26 and 20, respectively. Account for the differences.

4.4. Suggest plausible mechanisms for the following reactions, and comment on the reaction steps you suggest:

(a) $ClCH_2CH_2CH_2CN \xrightarrow{\text{NaNH}_2}$ ▷—CN + NaCl

(b) $CH_3COCH_2CO_2Et \xrightarrow[\text{(2) } I_2]{\text{(1) } ^-OEt}$ $\begin{array}{c} MeCOCHCO_2Et \\ | \\ MeCOCHCO_2Et \end{array}$

(c) $Et_3Si–H \xrightarrow[\text{CHCl}_3]{t\text{-BuO}^-} Et_3Si–CCl_2–H$

4.5. The benzoin condensation of benzaldehyde in the presence of the cyanide ion in aqueous solution is believed to have the following mechanism:

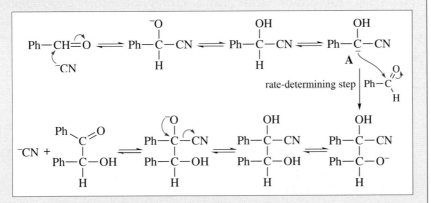

(a) What kinetics would be expected?

(b) By considering the stability of the carbanion **A**, provide an explanation for the fact that the reaction is not catalysed by the hydroxide ion, OH^-.

4.6. The keto alcohol $Me_2C(OH)CH_2COMe$ in aqueous sodium hydroxide is converted into acetone (propanone), $MeCOMe$. Suggest a plausible mechanism, and the kinetics expected.

5
Cations and Electrophiles

Many reactions in polar solvents involve the production of cations as intermediates. This chapter focuses on **electrophiles**. These are electron-deficient species that form a covalent bond with a reaction partner (the nucleophile) by accepting both electrons for the bond from the nucleophile (reaction 5.1). The terms **Lewis acid** and **Lewis base** are also used to describe electrophiles and nucleophiles, especially in the context of inorganic chemistry. Electrophiles can be positively charged or neutral. If positively charged, they are **cations**. Cations centred on carbon are **carbocations**. All proton acids are electrophiles; other electrophiles include the nitronium ion NO_2^+ and the bromine molecule. Some examples of reactions of electrophiles with nucleophiles (which can be negatively charged or neutral) are shown in reactions (5.2)–(5.5). Reactions (5.2) and (5.3) involve positively charged electrophiles, whereas the electrophiles in (5.4) and (5.5) are neutral.

$$E^+ + {:}Nu^- \longrightarrow E-Nu$$

electrophile nucleophile   covalent
Lewis acid   Lewis base   bond $\qquad$ (5.1)

Electrophiles are positively charged or neutral
Nucleophiles are negatively charged or neutral

$$Me_3C^+ \quad {}^-OH \longrightarrow Me_3C-OH \qquad (5.2)$$

$$(5.3)$$

$$Br-H \quad H_2C{=}CH_2 \longrightarrow Br^- + H_3C-CH_2^+ \qquad (5.4)$$

$$(5.5)$$

electrophile nucleophile

The *tert*-butyl (2-methylprop-2-yl) cation, which is the electrophile in reaction (5.2), has a vacant orbital which can accept the electron pair from the nucleophile without the need for any further movement of electrons. However, in many reactions a further electron pair movement is needed. The electrophilic proton in H_3O^+, HBr or H_2O (reactions 5.3–5.5) can only support one covalent bond, so as the nucleophile attacks with its electron pair, the bond from the proton to the oxygen or bromine atom has to break, with the electrons from that bond forming a lone pair on the oxygen or bromine atom.

Aims

By the end of this chapter, you should:

- Understand the meaning of the terms electrophile, cation, carbocation
- Know how cations are detected as reaction intermediates
- Know the principal reactions of carbocations
- Understand the importance of acid-catalysed reactions
- Be aware of the types of reactions that may involve cationic intermediates, and be able to postulate plausible reaction mechanisms involving cations

5.1 Formation of Carbocations

Carbocations can be formed in a number of ways, including heterolysis of a covalent bond and addition of an electrophile to a multiple bond.

5.1.1 Heterolysis of a Covalent Bond

Organic compounds with covalent bonds to electronegative elements may dissociate to form carbocations, especially if the cation is stabilized by the inductive effect, as in the *t*-butyl cation, or by resonance, as in the cumyl (2-phenylprop-2-yl) cation. This can happen slowly in polar solvents such as water. The S_N1 reaction of *t*-butyl halides is an example (reaction 5.6). The slow and rate-determining heterolysis of the halide is followed by a rapid reaction of the *t*-butyl cation with water to give the alcohol product.

$$Me_3C-Cl \xrightarrow[slow]{-Cl^-} Me_3C^+ \xrightarrow[fast]{H_2O} Me_3C-OH \qquad (5.6)$$

Loss of halide can be assisted by the presence of a Lewis acid such as antimony(V) fluoride (reaction 5.7). In this or other polar aprotic solvents the carbocation formed has no nucleophile to react with except for the

SbF_5Cl^- anion, which regenerates the original halide. In these solutions, concentrations of cations can be achieved which are high enough to allow study by NMR spectroscopy.

$$Me_2CH-Cl \,\, \diagup SbF_5 \longrightarrow Me_2CH^+ + SbF_3Cl^- \qquad (5.7)$$

5.1.2 Addition of an Electrophile to a Multiple Bond

The proton and other electrophiles can readily add to one end of a C=C or C≡C double or triple bond, and to the more electronegative atom in C=N, C≡N or C=O bonds. In addition to a C=C or C≡C bond, the direction of addition is to give the most stabilized cation. Alkyl groups stabilize a cation by their inductive and hyperconjugative effects (see Chapter 3); unsaturated and aromatic groups stabilize a cation by delocalization of the positive charge onto remote atoms. Propene reacts with HBr in the first step to give the secondary prop-2-yl cation (reaction 5.8), rather than the less stable primary prop-1-yl cation (reaction 5.9). Protonation of a carbonyl group (reaction 5.10) gives a delocalized cation **1**, with the charge spread between the carbon and the oxygen atoms. This reaction is reversible and forms the first step of many acid-catalysed reactions.

Electrophiles normally add to the end of a multiple bond which gives the more stable cation intermediate.

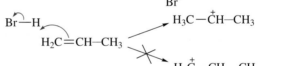

$$(5.8)$$

$$(5.9)$$

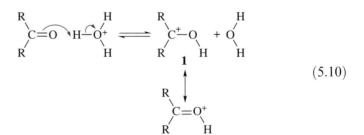

$$(5.10)$$

Addition of a proton or other electrophile is often rapidly followed by addition of a nucleophile at the other terminus of the multiple bond.

5.2   Evidence for Cations

5.2.1   NMR Spectroscopy

Reactions such as S_N1 reactions in water are believed to involve cationic intermediates, but do not give sufficient concentrations of the cations for them to be identified spectroscopically. However, cations can be observed conveniently in solutions in aprotic polar solvents such as SbF_5. The strong deshielding, resulting in high values of chemical shift for both ^{13}C and ^1H NMR spectra, is good evidence for cationic character. For example, the ^1H NMR spectrum of the prop-2-yl cation $(CH_3)_2CH^+$ shows a one-proton resonance at $\delta = 13.5$ ppm, split into septets by the six methyl protons. The methyl protons appear at 5.1 ppm. The very high value for the CH proton is consistent with its attachment to a carbon bearing a positive charge; the high value for the methyl protons suggests that the charge on the neighbouring carbon is causing deshielding by inductively and hyperconjugatively withdrawing electrons. The diphenyl-methyl cation **2** (Figure 5.1) shows a high-field proton at 9.8 ppm; the ring protons are more deshielded at the *ortho* and *para* positions, corresponding to greater positive charge density at these positions compared with the *meta* protons, in accord with resonance and molecular orbital theory.

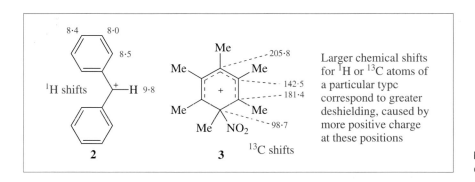

Larger chemical shifts for ^1H or ^{13}C atoms of a particular type correspond to greater deshielding, caused by more positive charge at these positions

Figure 5.1 ^1H and ^{13}C NMR chemical shifts (δ)

The nitronium ion NO_2^+ attacks hexamethylbenzene to give an adduct **3** of the type thought to be involved in electrophilic aromatic substitution; the ^{13}C chemical shifts for the ring carbon atoms are shown in Figure 5.1. The δ value of 98.7 ppm is very high for a sp^3 carbon atom and is attributed to the positive charge build-up on the adjacent carbon atoms. Again the *ortho* and *para* carbon chemical shifts are higher than those of the *meta* carbon atoms, in line with resonance and molecular orbital theory.

Carbocations show characteristic ^1H and ^{13}C NMR spectra with chemical shifts for the charged centres downfield of all neutral centres, due to the deshielding effect of the positive charge.

Worked Problem 5.1

Q Draw resonance structures for **2** and **3** to show the positions at which positive charge build-up is expected.

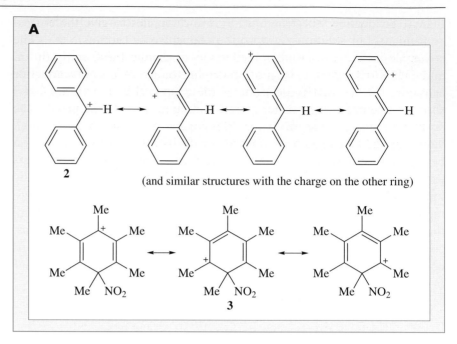

5.2.2 Trapping

Cations react rapidly and indiscriminately with nucleophiles. Thus if a reaction which is suspected to go *via* a cationic intermediate is carried out in the presence of an added nucleophile, and an adduct containing the new nucleophile is obtained, this provides evidence for a cationic intermediate which is trapped by the added nucleophile. For example, the addition of bromine to alkenes is thought to go *via* a cationic intermediate **4** (reaction 5.11). If chloride or nitrite ions are added to the reaction mixture, the chloride or nitrite adducts **5** and **6** are obtained, even though the chloride and nitrite ions do not react with ethene (or 1,2-dibromoethane) directly at a rate that would account for the amount of these products in the reaction mixture. This provides strong evidence for the intermediate bromoethyl cation **4**, which will be trapped by the added chloride or nitrite ions (reactions 5.12 and 5.13). The structure of bromoalkyl cations is discussed further in the section on electrophilic addition.

$$Br-CH_2-CH_2-Br \qquad (5.11)$$

$$Br-Br$$
$$H_2C=CH_2 \longrightarrow H_2\overset{|}{C}-\overset{+}{C}H_2 \xrightarrow{\quad Cl^- \quad} Br-CH_2-CH_2-Cl \qquad (5.12)$$
$$\underset{\mathbf{4}}{} \qquad\qquad \underset{\mathbf{5}}{}$$

$$\xrightarrow{\quad NO_2^- \quad} Br-CH_2-CH_2-O-N=O \qquad (5.13)$$
$$\underset{\mathbf{6}}{}$$

Worked Problem 5.2

Q Show that attack by Br⁻ at the 3-position of **30** would give the same product.

A

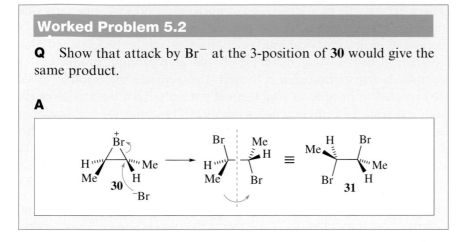

Epoxidation by Peroxy Acids

The O–O bond in peroxycarboxylic acids **32** can be regarded as polarized in the direction Ar–CO–O$^{\delta-}$–O$^{\delta+}$H, so **32** reacts by electrophilic addition of the O$^{\delta+}$H group to a double bond (reaction 5.23). However, a lone pair of electrons on the oxygen atom immediately coordinates with the other carbon which would otherwise have developed the positive charge, to give the bridged intermediate hydroxonium ion **33**. The benzoate anion escapes and takes no further part in the reaction; the stabilization of the benzoate anion provides a driving force for the reaction. The oxonium ion **33** is attacked from the opposite side by a water molecule in an S$_N$2 reaction to form the cation **34**. This loses a proton to give the diol **35**, in which the two hydroxyl groups have been added to the original C=C bond in the *anti* (opposite sides) orientation.

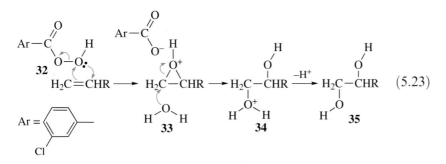

(5.23)

S$_N$1 Reactions

S$_N$1 reactions (5.24), discussed in Chapter 1, involve a slow rate-determining step in which a leaving group departs, taking the electrons from the covalent bond, and leaving a reactive carbocation intermediate.

This reacts rapidly by addition to any nucleophile present, especially with the solvent, to form a stable product. A final proton loss may be necessary. Since carbocations are planar, it might be expected that optically active starting materials should give racemic products: if a completely free carbocation is formed, the incoming nucleophile would be expected to be able to attack equally readily from both sides of the cation (Figure 5.2a). In practice, a mixture of racemization and inversion is observed, suggesting that the leaving group is still in the vicinity, either free or as an ion pair, which inhibits attack from the same side as the departing group and therefore favours attack from the opposite side, giving rise to a greater amount of inversion than retention (Figure 5.2b).

$$R\text{—Hal} \xrightarrow{\text{slow}} Hal^- + R^+ \xrightarrow{H_2O} R\text{—}\overset{+}{O}H_2 \xrightarrow{H_2O} R\text{—OH} + H_3O^+ \qquad (5.24)$$

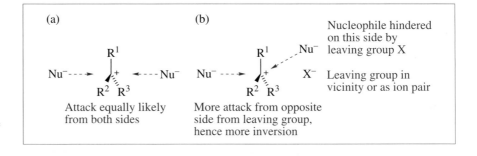

Figure 5.2 Stereochemistry of S_N1 reactions

5.3.3 Elimination

The alternative way in which a carbocation can be converted into a neutral product is by elimination of a cation, almost always a proton. Two important reactions involve elimination as the final step.

Unimolecular Elimination (E1) Reactions

When a carbocation is formed by loss of a nucleophile, as in the S_N1 reaction, elimination of a proton will always compete with substitution, provided that there is a β-C–H bond in the cation. These eliminations are described as **unimolecular eliminations (E1)** because, as with the competing S_N1 reaction, the rate-determining step is the unimolecular reaction of the original halide to give the carbocation. For example, *t*-butyl chloride in ethanol gives the *t*-butyl cation, which can either react with a solvent molecule to give the ethyl ether **36** (S_N1, **solvolysis**, reaction 5.25) or a proton can be removed by a solvent molecule to give 2-methylpropene **37** (E1, reaction 5.26). If more than one type of β-C–H bond is available, a mixture of products will be formed. The proportion of products is mainly determined by the stability of the alkene produced. For unhindered monoalkenes, this will favour the alkene product which has most substituents on the C=C, since this alkene will be the most stable. For example, $Me_2C=CHMe$ ($\Delta H_f = -42.1$ kJ mol^{-1}, three substituents) is more stable than $Me(Et)C=CH_2$ ($\Delta H_f = -35.6$ kJ mol^{-1}, two substituents), which in turn is more stable than $Me_2CH-CH=CH_2$ ($\Delta H_f = -27.4$ kJ mol^{-1}, one substituent). This stabilization is probably due to the greater strength of the $C_{sp2}-C_{sp3}$ bond than the $C_{sp3}-C_{sp3}$ single bond, due to better overlap. There are more of these bonds in more substituted alkenes. However, bulky groups such as the *t*-butyl group cause steric destabilization, especially when *cis* to another group at the other end of the double bond. Examples of these effects are shown in reactions (5.27) and (5.28). In (5.27), the trisubstituted alkene **38** predominates over the disubstituted alkene **39**; in (5.28), the predominant product is the disubstituted isomer **40** where the *t*-butyl and methyl groups are *trans* to each other, followed by the monosubstituted **41**, with a very small amount of the *cis* product **42**, which, although disubstituted, has significant steric hindrance between the two groups *cis* to each other. The proportions follow the measured heats of formation of the three isomers.

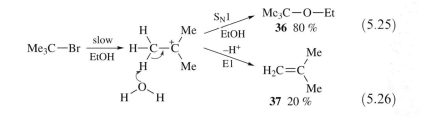

(5.25)

(5.26)

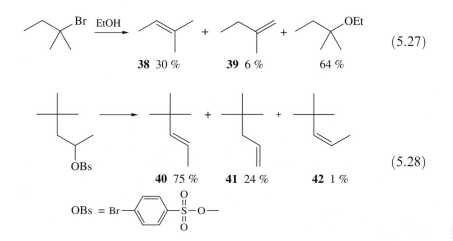

(5.27)

(5.28)

Elimination from halides almost always gives a mixture of elimination and solvolysis products. The acid-catalysed elimination of water from alcohols provides a preparative alternative (reaction 5.29). The protonated alcohol **43** loses a water molecule to give the carbocation, which can eliminate a proton to form the alkene. Even if some of the carbocation is trapped as the sulfate ester **44**, this reaction is reversible, and the alkene can be distilled out of the reaction mixture to bring the reaction to completion.

$$
\text{Me}_3\text{C}-\text{OH} \xrightarrow[\text{H}_2\text{SO}_4]{\text{H}^+} \underset{\textbf{43}}{\text{Me}_3\text{C}-\overset{+}{\text{O}}\text{H}_2} \xrightarrow{-\text{H}_2\text{O}} \text{Me}_3\text{C}^+ \xrightarrow{-\text{H}^+} \text{alkene}
$$

$$
\underset{\textbf{44}}{\text{Me}_3\text{C}-\text{OSO}_3\text{H}} \rightleftharpoons \text{Me}_3\text{C}-\overset{H}{\underset{SO_3H}{\overset{+}{O}}}
$$

(5.29)

Electrophilic Aromatic Substitution

In these important reactions, one electrophile replaces another (usually a proton) on an aromatic ring. In a classical example, benzene is converted into nitrobenzene by a mixture of nitric and sulfuric acids (reaction 5.30). The reagent that attacks the benzene ring is believed to be the nitronium ion, NO_2^+, for which there is spectroscopic and other evidence. Two possible mechanisms are shown in reactions (5.31, $\text{E}^+ = \text{NO}_2^+$). The first (route a) would involve simultaneous bond making and bond breaking, going through a transition state **45** to the substituted product **47**. Evidence against this route is the absence of a hydrogen isotope effect, which shows that the C–H bond is not significantly broken in the transition state, and the effect of substituents on the benzene ring. The rate of reaction is speeded up by electron-releasing substituents such as methyl, supporting the presence of a positively charged intermediate or transition state. Additionally, the effect of substituents at the *para* position on the rate shows that the effect correlates with σ^+ rather than σ (see Chapter 3), which shows that the charge is conjugated with the *para* substituent, allowing extra stabilization as indicated in structures **49** and **50**. The involvement of an intermediate **46** (a Wheland intermediate) and thus route (b) is supported by NMR evidence (see the section on structure of carbocations above), which shows that structures of this type can be identified for highly substituted benzenes. Salts of cations of type **46** have recently been isolated as crystals whose structures have been determined by X-ray crystallography.

$$HNO_3 + H_2SO_4 \longrightarrow NO_2^+ \xrightarrow{\ C_6H_6\ } C_6H_5NO_2 + H^+ \qquad (5.30)$$

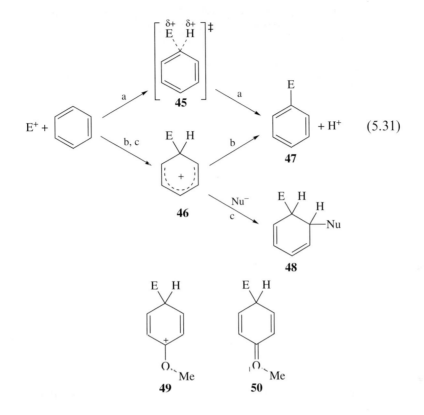

$$(5.31)$$

Why do the adduct carbocations **46** not react with a nucleophile to give an adduct of type **48**, as happens in electrophilic addition to double bonds? The main reason is probably thermodynamic. The hypothetical reaction (5.32) is very exothermic, based on experimental heats of formation. If we attribute this exothermicity to the stabilization of the benzene nucleus, we obtain a value of 150 kJ mol^{-1} for this stabilization. This stabilization would be lost in reaction (5.31) if the adduct **48** were formed, but is regained if the substituted product **47** is produced. This stabilization does not apply to reaction with alkenes, so addition can take place if the reaction is favourable energetically.

Electrophiles add to both alkenes and aromatic compounds to give an intermediate adduct cation. For alkenes, this cation normally reacts with a nucleophile to give an adduct, rather than eliminating another electrophile. For aromatic compounds, elimination of an electrophile regains the original aromatic stabilization energy, so this is the preferred route.

$$3 \ \bigcirc \longrightarrow \bigcirc + 2 \ \bigcirc \qquad \Delta H = -150 \text{ kJ mol}^{-1} \qquad (5.32)$$

5.4   Electrophilic Rearrangements involving Migration of C to O or N

Rearrangements of carbocations by migration of a carbon-centred group from an adjacent atom as in reaction (5.33) are common, and this is indeed a characteristic reaction, as discussed in Section 5.3.1. We might expect that similar reactions might be observed for positively charged nitrogen and oxygen species with six electrons in their valence shell, as in reactions (5.34) and (5.35). However, the greater electronegativity of nitrogen and oxygen makes species such as **51** and **53** so unstable that it is almost impossible to create them in solution. Thus in reaction (5.36) the first stage involves the protonation of the oxygen atom in the oxime **55** to give **56** in which the OH_2 fragment would be a good leaving group, but its loss with a pair of electrons to leave the positively charged nitrogen-centred intermediate **57** is too endothermic to take place. Instead, the production of **57** can be forestalled by the simultaneous migration of the group R^1 to the nitrogen atom to give the carbocation **58**, which in turn is attacked by a water molecule to give **59**, and after three subsequent protonation and deprotonation steps gives the amide **60** in good yield. Evidence that the reaction is concerted and does not involve **57** as an intermediate is provided by the observation that the isomeric oxime **61** reacts to give the amide **62** in which R^2, the other group, has migrated. If the symmetrical intermediate **57** were involved, a mixture of amides **60** and **62** would be expected from the rearrangement of either of the oximes **55** or **61**. This rearrangement is useful synthetically and is known as the **Beckmann rearrangement**.

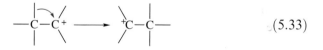

(5.33)

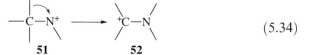

(5.34)

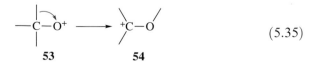

(5.35)

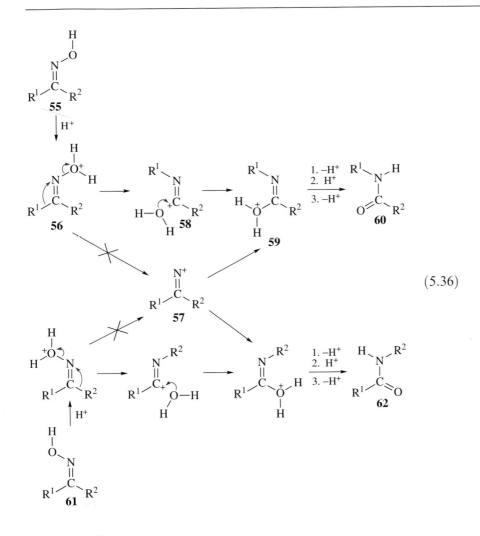

(5.36)

Ketones such as **63** are oxidized by peroxy acids **65** in the **Baeyer–Villiger reaction** (5.37), which involves migration of a carbon-centred group from carbon to oxygen. The protonated ketone **64** reacts with the peroxy acid **65** to give the ester **66**. Migration of R^2 to oxygen takes place as the stabilized carboxylate anion **68** leaves, to give the protonated ester **67**. This loses a proton to give the ester product **70**. The alternative route via the unstable oxygen-centred cation **69** does not take place.

In the Beckmann and Baeyer–Villiger rearrangements, migration of a carbon-centred group from C to N or O takes place synchronously with elimination of a good leaving group. If the leaving group left first, very unstable R_2N^+ or RO^+ intermediates would be formed.

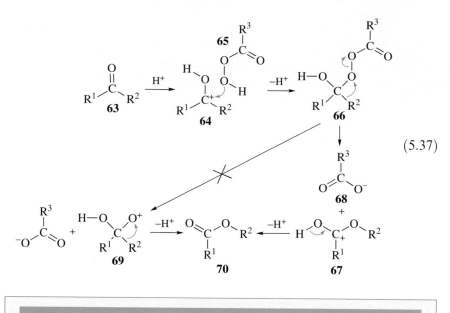

(5.37)

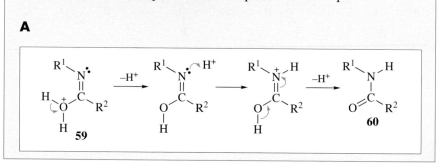

Worked Problem 5.3

Q Show how the intermediate **59** (reaction 5.36) is converted into the amide **60** in three protonation–deprotonation steps.

A

5.5   Acid-catalysed Reactions

Many reactions are catalysed by acids. Hydrolysis of esters and the reverse reaction, esterification, are important examples both in the laboratory and in living systems (reaction 5.38). In the forward direction, the initial rate is proportional to $[H_3O^+][71]$; in the reverse direction the kinetic dependence is on $[H_3O^+][74][EtOH]$. By the principle of microscopic reversibility (Chapter 1), the reaction must have the same mechanism in both directions. The third-order kinetic dependence of the reverse reaction suggests that the reaction is complex and involves intermediates, because termolecular reactions are rare. The formation of tetrahedral intermediates of type **72** or **73**, involving reversible addition of a water molecule, is supported by the observation that if partial hydrolysis of an ester labelled with ^{18}O in the carbonyl group is carried out and unchanged ester is recovered, oxygen exchange with the solvent water occurs. In both reaction directions, if substituted benzoate esters are used, negative Hammett ρ factors are observed, showing that positive charge is built up during the reaction, and the dependence on σ^+ shows that the charge is conjugated with the ring (see Section 3.3 for an explanation of substituent effects).

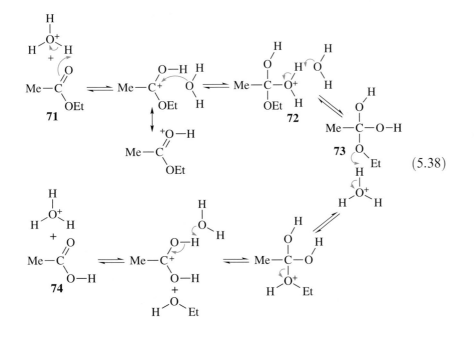

(5.38)

The scheme shown in (5.38) is consistent with these observations. Why are so many steps involved? The scheme has been drawn out in detail, including the representation of the solvated proton as H_3O^+ to emphasize that *in each step a bond is formed and a bond is broken. Thus all the steps are approximately thermoneutral, and can take place with low activation energies. The number of steps is immaterial, so long as there is a negative Gibbs free-energy change in the reaction.* Most biological reactions involve sequences of this type.

In general, the fact that a reaction is acid catalysed implies that a substrate molecule is protonated in the first step, and that the protonated molecule reacts further, perhaps with other molecules, to give product molecule(s), regenerating the proton during the process. This route must be faster than alternative schemes which do not involve the catalyst, otherwise no catalysis would have taken place. Two possibilities are shown in reactions (5.39) and (5.40).

$$\text{S} + \text{H}^+ \underset{}{\overset{\substack{\text{rapid} \\ \text{pre-equilibrium}}}{\rightleftarrows}} \text{SH}^+ \xrightarrow{\text{slow}} \text{Products} \qquad (5.39)$$

$$\text{S} + \text{H}^+ \underset{\substack{\text{reverse deprotonation} \\ \text{does not compete with} \\ \text{product formation}}}{\overset{}{\rightleftarrows}} \text{SH}^+ \xrightarrow{\text{fast}} \text{Products} \qquad (5.40)$$

In (5.39), a rapid pre-equilibrium is set up between the substrate S and its protonated form SH^+, followed by a slow reaction of SH^+ to give products. Alternatively, the protonation may be rate determining, with a rapid reaction of SH^+ to give products being faster than the reverse deprotonation back to S, as in reaction (5.40). There are kinetic consequences of this difference. For reactions of type (5.39), the set-up of an equilibrium between S and SH^+ means that the concentration of SH^+ and thus the rate depends only on the (solvated) hydrogen ion concentration. Reactions of this type are said to undergo **specific acid catalysis**; the rate will depend only on the concentration of solvated protons and not on the concentration of any other acid species present, including water. However, for reactions of type (5.40), where protonation is rate determining, the protonation can be carried out either by the solvated proton (reaction 5.41) or by any other acid present, including water (reaction 5.42). Each route will independently produce SH^+ ions that contribute to product formation. The rate equation will have terms involving all acids present, and the reaction is said to be subject to **general acid catalysis**.

$$\text{S} + \text{H}_3\text{O}^+ \longrightarrow \text{SH}^+ + \text{H}_2\text{O} \qquad (5.41)$$

$$\text{S} + \text{HA} \longrightarrow \text{SH}^+ + \text{A}^- \qquad (5.42)$$

An example of specific acid catalysis is the hydrolysis of acetaldehyde diethyl acetal (1,1-diethoxyethane, **75**). Reaction (5.43) was carried out in a formic (methanoic) acid/formate buffer. The formic acid concentration was varied over a factor of 10, whereas the ratio $[HCO_2H]/[HCO_2^-]$ was maintained at a value of 3.0, to give a constant pH and hydrogen ion concentration. The rate of hydrolysis in all the experiments was the same, within experimental error, showing that the rate depended only on the hydrogen ion concentration and not on the formic acid or water concentrations. Thus this reaction is subject to specific acid catalysis. A rapid pre-equilibrium is set up between **75** and its protonated form **76** in the first step. The second and rate-determining step involving loss of an ethanol molecule to give the carbocation **77** is slow, so specific acid catalysis by the solvated proton is observed. The subsequent steps lead to an acetaldehyde (ethanal) molecule **78** in good yield.

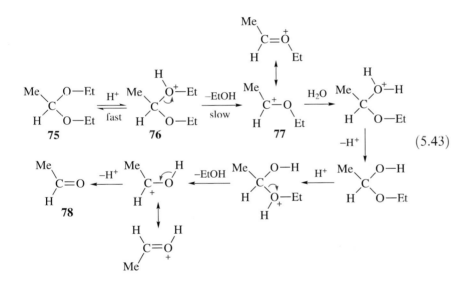

(5.43)

An example of specific acid catalysis.

By contrast, the hydrolysis of 1,1,1-triethoxyethane (ethyl ortho-acetate, **79**) to acetic acid (**82**) is subject to general acid catalysis (reaction 5.44), as discussed in Chapter 2 (reaction 2.32). When the reaction is carried out in a *m*-nitrophenol/*m*-nitrophenolate buffer, terms in $[H_3O]^+$, $[m\text{-}O_2NC_6H_4OH]$ and $[H_2O]$ are found in the rate equation. In reaction (5.44), the first step to give the protonated compound **80** is rate determining, followed by a rapid second step.

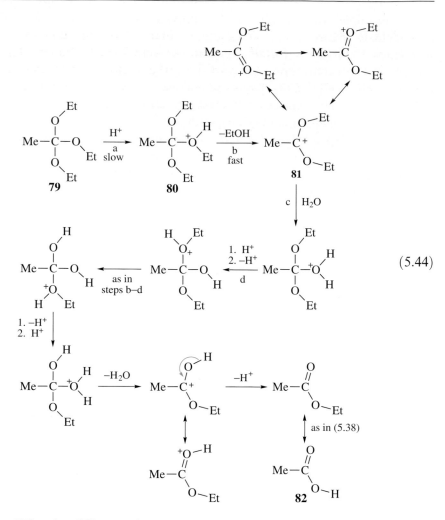

An example of general acid catalysis.

(5.44)

In acid-catalysed reactions, the first step involves the protonation of a neutral substrate molecule S. If the protonation is rate-determining, general acid catalysis will be observed. If a rapid pre-equilibrium between S and SH+ is followed by a later rate-determining step, specific acid catalysis will take place.

Why the difference between (5.43) and (5.44)? In reaction (5.43), the carbocation **77** formed in the second step is stabilized by only one oxygen substituent. However, the analogous cation **81** in reaction (5.44) is stabilized by two oxygen substituents, making the fragmentation step more exothermic and therefore faster, so in this case the rate-determining step shifts from the second to the first.

Summary of Key Points

1. Electrophiles react with nucleophiles by accepting an electron pair from the nucleophile to form a covalent bond. Carbocations and solvated protons are strongly electrophilic; many neutral species such as hydrogen halides and halogen molecules are also electrophilic.

2. Important reactions of carbocations are rearrangement, addition to multiple bonds, and elimination. Rearrangements are much more common in carbocations than in carbanions or free radicals; if a carbon skeleton rearrangement has taken place in a chemical reaction, it is likely that a carbocation intermediate is involved.

3. A number of rearrangements of oxygen- or nitrogen-containing organic compounds involve a concerted migration of a carbon-centred group from carbon to oxygen or nitrogen to avoid the production of the very unstable intermediates of type RO^+ or R_2N^+.

4. Many reactions are catalysed by acids; most involve a reversible protonation of a neutral substrate as the first step. In specific acid catalysis the kinetics show a single rate constant which involves the concentration of the solvated proton and one or more substrate molecules. In general acid catalysis, the observed rate involves several terms, each with a different rate constant, one for each acid present in the reaction mixture.

5. Acid-catalysed reactions often involve many steps, in each of which a bond is made and a bond is broken in a concerted manner so that each reaction step is approximately thermoneutral and no high activation energy steps are involved. In these reactions, the intermediates and reagents are either neutral or positively charged. No anions are involved.

Further Reading

T. H. Lowry and K. S. Richardson, *Mechanism and Theory in Organic Chemistry*, HarperCollins, London, 1987.

H. Maskill, *Mechanisms of Organic Reactions*, Oxford University Press, Oxford, 1996.

C. A. Reed, K. -C. Kim, E. S. Stoyanov, D. Stasko, F. S. Tham, L. J. Mueller and P. W. D. Boyd, *Isolating benzenonium ion salts*, in *J. Am. Chem. Soc.*, 2003, **125**, 1796.

Problems

5.1. Which of the following species are (a) electrophiles, (b) nucleophiles, (c) carbocations, (d) cations. Some species may fit into more than one category, or into none:

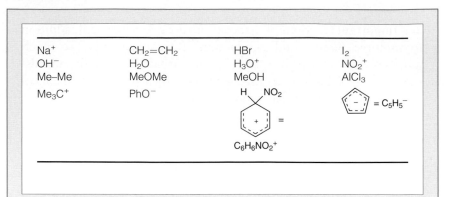

5.2. Would you expect Me_3CBr or $Me_2(Ph)CBr$ to react more rapidly in S_N1 reactions? Why?

5.3. Explain why although the S_N1 hydrolysis (a) is rapid, the analogous reaction (b) is very slow.

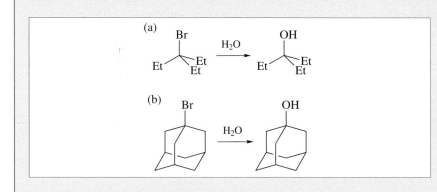

5.4. $PhCHMe–CHMeNH_2$ reacts with nitrous acid (HNO_2) in acetic acid to give a mixture of $PhCHMe–CHMeOAc$ (44%), $PhCMe(OAc)–CH_2Me$ (24%) and $PhCH(OAc)–CHMe_2$ (32%) [$Ac=CH_3CO$]. It is thought that one of the following is involved as an intermediate: $PhCHMe\text{-}CHMe^+$, $PhCHMe\text{-}CHMe^\bullet$ or $PhCHMe–CHMe^-$. Which is the most likely, and why?

5.5. 1-Fluoropropane in SbF_5 shows a ^1H NMR spectrum consisting of a doublet ($J = 3$ Hz) at $\delta = 5.1$ ppm and a septet ($J = 3$ Hz) at $\delta = 13.5$ ppm, with relative intensities of 6:1. Explain the chemistry that gives rise to the species showing this NMR spectrum, and comment on the chemical shifts observed.

5.6. One of the isomeric oximes **A** rearranges in the presence of acid to give the amide **B**. In the presence of base, the oxime undergoes ring closure to give **C**. Suggest mechanisms for both these reactions and identify which of the stereoisomeric oximes corresponds to **A**. What information about the stereochemistry of the migration of the methyl group in the rearrangement of **A** to **B** is provided by this experiment?

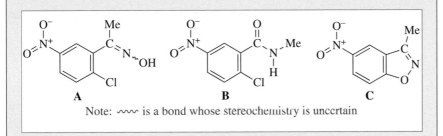

Note: ∿ is a bond whose stereochemistry is uncertain

5.7. The ester **D** is hydrolysed in H_2O/HOAc to give **F**, and it has been suggested that **E** is an intermediate. Devise a plausible mechanism for the reaction which involves **E** on the route and comment on any points of interest. How could you establish if the reaction is acid catalysed, and, if so, whether the catalysis is general or specific? Suggest a trapping experiment to support the intermediacy of **E**.

Note the change in stereochemistry here

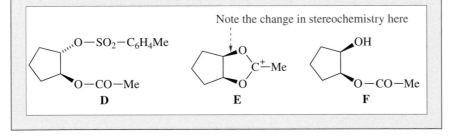

6

Radicals

Free radicals are species that have an unpaired electron in their valence shell. They tend to be very reactive and are normally only present in tiny concentrations. In solution in non-polar solvents, free radical and molecular reactions play a dominant role to the virtual exclusion of cations and anions. Almost all the reactions that take place in the atmosphere are free radical processes.

Aims

This chapter describes the most common types of radical reaction that occur in the gas phase or in solution, especially in non-polar solvents. By the end of this chapter you should be able to:

- Apply tests to organic reactions to confirm or refute the presence of free radical intermediates
- Understand the basic principles of electron spin resonance and the information that this technique provides about free radical structure
- Understand the importance of free radical chain reactions
- Know the key types of elementary reaction
- Suggest likely mechanisms for reactions known to involve free radical intermediates

6.1   Formation of Free Radicals

Since most organic molecules have all their electrons paired, conversion into radicals involves homolysis of a covalent bond, or a single electron transfer to or from a molecule.

6.1.1 Thermolysis

For molecules with a weak covalent bond, heating provides enough thermal energy to break the covalent bond homolytically (one electron from the covalent bond to each of the two radicals formed); the weaker the bond, the lower the temperature for decomposition. In practice, peroxides and azo compounds are the most practical and cleanest sources, as exemplified by reactions (6.1) and (6.2). Di-*t*-butyl peroxide (**1**) decomposes at temperatures above 100 °C to give two *t*-butoxyl radicals (**2**). The oxygen-oxygen bond is by far the weakest in the molecule, so this bond is the exclusive bond broken in the homolysis, giving a clean reaction. In azobisisobutyronitrile (AIBN, **3**) the C–N bond is the weakest, partly because the product 2-cyanoprop-2-yl radical (**4**) is stabilized by delocalization. The intermediate azo radical **5** quickly loses a nitrogen molecule to give a second 2-cyanoprop-2-yl radical, so reaction (6.2) provides a clean source of these radicals at temperatures above about 80 °C. Hydrocarbons such as 1,2-diphenylethane (**6**), which homolyse to give two stabilized benzyl radicals **7** (reaction 6.3), decompose at temperatures of about 400 °C, too hot to be of practical value for reactions in solution, whereas simple alkanes such as ethane only undergo homolysis at temperatures above 800 °C, making these reactions significant only in high-temperature pyrolyses and combustion reactions (*e.g.* reaction 6.4).

> Free radicals have an unpaired electron in the valence shell. They are formed by homolysis of a covalent bond, or by electron gain or loss from molecules which contain only paired electrons.

$$Me_3C\text{--}O\text{--}O\text{--}CMe_3 \longrightarrow Me_3C\text{--}O^\bullet + {}^\bullet O\text{--}CMe_3 \qquad (6.1)$$
$$\mathbf{1} \qquad\qquad\qquad \mathbf{2} \qquad\qquad \mathbf{2}$$

$$(6.2)$$

$$Ph\text{--}CH_2\text{--}CH_2\text{--}Ph \longrightarrow Ph\text{--}CH_2^\bullet + {}^\bullet CH_2\text{--}Ph \qquad (6.3)$$
$$\mathbf{6} \qquad\qquad \mathbf{7} \qquad \mathbf{7}$$

$$CH_3\text{--}CH_3 \longrightarrow 2\ CH_3^\bullet \qquad (6.4)$$

6.1.2 Photolysis

The energy required to break a covalent bond can also be provided by electromagnetic radiation, normally ultraviolet or visible light.

The molecule needs to have a chromophore to absorb the radiation. Chlorine, bromine and iodine molecules absorb visible radiation and are decomposed by a photon of light into two halogen atoms (reaction 6.5). Alkanes do not absorb visible or UV light down to 200 nm, and thus do not undergo photolysis to free radicals. Ketones such as benzophenone (diphenylmethanone, **8**) absorb near-UV light to give an excited molecule **9** in which an electron has been promoted from a non-bonding orbital on oxygen to the π^* antibonding orbital of the carbonyl group. This n→π^* transition gives an excited ketone molecule **9**, which has enough energy to abstract a hydrogen atom from a solvent (usually alcohol) molecule to give the stabilized $Ph_2\overset{\bullet}{C}OH$ radical **10** and a more reactive alcohol-derived radical $R_2\overset{\bullet}{C}OH$ **11**, both of which can undergo further reactions (reaction 6.6 and Figure 6.1).

$$X{-}X \xrightarrow{\ hv\ } X^{\bullet} + X^{\bullet} \tag{6.5}$$
$$X = \text{Cl, Br or I}$$

$$Ph_2C{=}O \xrightarrow{\ hv\ } Ph_2\overset{\bullet}{C}{-}\overset{\bullet}{O} \xrightarrow{\ HCR_2OH\ } Ph_2\overset{\bullet}{C}{-}OH + \overset{\bullet}{C}R_2OH \tag{6.6}$$
$$\textbf{8} \qquad\qquad \textbf{9} \qquad\qquad\qquad \textbf{10} \qquad \textbf{11}$$

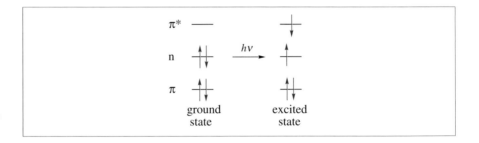

Figure 6.1 n→π^* excitation of ketones

6.1.3 Electron Transfer

Many metal atoms can give up a single electron to give a cation with a single positive charge, and many ions can give up or take an electron from an organic molecule to convert it into a radical. Practical examples are shown in reactions (6.7)–(6.9). In (6.7) the silver atom gives up an electron to the chlorotriphenylmethane molecule **12**, which promptly loses a chloride ion to give the triphenylmethyl radical **13**, whereas in (6.8) the electron is transferred from the sodium atom to the lowest antibonding orbital of anthracene (**14**) giving a radical anion **15** which behaves both as a radical and as an anion. Fe(II) salts are readily oxidized to Fe(III) salts by hydrogen peroxide, giving a hydroxide ion and a hydroxyl radical, a convenient source of the latter at room temperature (reaction 6.9).

$$\text{Ph}_3\text{C–Cl} + \text{Ag} \longrightarrow \text{Ph}_3\text{C}^\bullet + \text{Ag}^+\text{Cl}^- \qquad (6.7)$$
$$\quad\;\; \mathbf{12} \qquad\qquad\quad \mathbf{13}$$

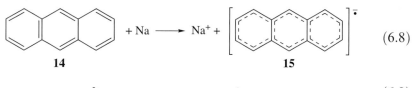

$$\mathbf{14} \qquad\qquad\qquad\qquad\qquad \mathbf{15} \qquad (6.8)$$

$$\text{Fe}^{2+} + \text{H–O–O–H} \longrightarrow \text{Fe}^{3+} + \text{HO}^\bullet + {}^-\text{OH} \qquad (6.9)$$

6.1.4 Molecule-induced Homolysis

Some reactions occur thermally at temperatures too low for homolysis of any of the covalent bonds present to provide enough radicals to start the reaction. For example, mixtures of fluorine and methane explode at room temperature, and many hydrocarbons are oxidized slowly by molecular oxygen (see below). It has been postulated that the bimolecular reactions (6.10) and (6.11) are responsible. In (6.10), the formation of the very strong H–F bond makes this reaction much less endothermic than the simple homolysis of the fluorine molecule. In (6.11), a strong O–H bond is formed in the hydroperoxyl radical **18**, whereas two relatively weak bonds are broken, the O=O π bond and the C–H bond in **16** which is weakened by the stabilization of the product benzylic radical **17**. The occurrence of these molecule-induced homolysis reactions is difficult to prove because the compounds formed tend to be swamped by those from the subsequent radical reactions.

$$\text{CH}_4 + \text{F–F} \longrightarrow \text{CH}_3^\bullet + \text{H–F} + \text{F}^\bullet \qquad (6.10)$$

$$\text{PhCMe}_2\text{H} + \text{O=O} \longrightarrow \text{Ph}\overset{\bullet}{\text{C}}\text{Me}_2 + \text{H–O–O}^\bullet \qquad (6.11)$$
$$\qquad\; \mathbf{16} \qquad\qquad\qquad\quad \mathbf{17} \qquad\quad \mathbf{18}$$

6.2   Destruction of Radicals (Termination)

Radicals show a very important difference from anionic and cationic intermediates. Because radicals are uncharged and possess an unpaired electron in an orbital which is capable of forming a covalent bond, they normally react with each other on every collision with little or no activation energy. The bimolecular rate constants are large, in the region of 10^{10} M^{-1} s^{-1}. **Combination** results in a new covalent bond, as in reaction (6.12). If more than one type of radical is present in the system, cross-combination occurs, for example reaction (6.13), and the final product contains all possible dimers in proportions that depend on the relative concentrations of the radicals. For alkyl radicals such as the *t*-butyl radical **19** with one or more β-hydrogen atoms, an alternative reaction is disproportionation, in which one radical abstracts a hydrogen atom from the β-position of the other, leading to an alkane and an alkene product (reaction 6.14). The more β-hydrogen atoms present, the more does disproportionation compete with combination.

$$CH_3^{\bullet} \quad {}^{\bullet}CH_3 \longrightarrow CH_3–CH_3 \tag{6.12}$$

$$CH_3^{\bullet} + Cl^{\bullet} \longrightarrow CH_3–Cl \tag{6.13}$$

$$Me_3C^{\bullet} \quad H–CH_2–\overset{\displaystyle Me}{\underset{\displaystyle Me}{\overset{|}{\underset{|}{C^{\bullet}}}}} \longrightarrow Me_3CH + CH_2{=}\overset{\displaystyle Me}{\underset{\displaystyle Me}{\overset{/}{\underset{\backslash}{C}}}} \tag{6.14}$$

$$\quad \textbf{19} \qquad\qquad\qquad \textbf{19}$$

The rapid rate of radical termination reactions for simple organic radicals means that only very small steady-state concentrations can be maintained, typically 10^{-8} M or lower. However, some larger stabilized radicals do not dimerize readily and therefore larger concentrations can be built up. These stabilized radicals will be considered later in the chapter.

6.3   Detection of Radicals as Reaction Intermediates

Because of the low steady-state concentrations involved, this is not always easy. Evidence for free radical intermediates can be obtained from:

Kinetics. As outlined in Chapter 2, since radicals tend to be formed by first-order processes and destroyed in second-order reactions, steady-state radical concentrations are usually proportional to the square root of the precursor concentration [A]. This often leads to kinetic dependence on $[A]^{0.5}$ or $[A]^{1.5}$, so kinetic dependence of this form is strong evidence for a free radical mechanism.

Influence of light. If a reaction is speeded up by visible or UV light, a free radical mechanism should be considered, though we shall see in the next chapter a variety of molecular photo-induced reactions are also possible. The quantum yield Φ (phi) is a useful parameter, defined as the number of molecules converted from starting materials into products per photon absorbed. A few radical reactions have Φ in the region of 1 or 2, but most radical chain reactions have higher Φ values, often 1000 or more, so a high quantum yields for any reaction taking place in the gas phase or in solution in non-polar solvents is a strong indication of a radical reaction.

Initiation and inhibition. If a reaction is speeded up by the addition of compounds such as peroxides which are known to produce free radicals, this is evidence for a free radical mechanism. Inhibition (slowing down) of a reaction by compounds such as phenols is also evidence for free radical behaviour. 2,4,6-Tri-*t*-butylphenol (**20**) acts as an inhibitor by hydrogen transfer (reaction 6.15) to one of the chain-carrying radicals involved, giving the phenoxyl radical **21**, which is stabilized by delocalization and may be unable to carry on the chain reaction.

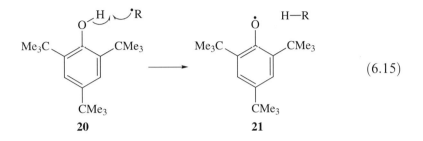

$$(6.15)$$

Initiation of polymerization. If styrene (phenylethene, $PhCH=CH_2$) or methyl methacrylate (methyl 2-methylpropenoate, $CH_2=C(Me)-CO_2Me$),or preferably an equimolar mixture of the two, is added to the reaction mixture, the production of polymer, which can be precipitated from the mixture by methanol, indicates the presence of radicals. If a mixture of the alkene monomers is used, the product will be a copolymer containing equal quantities of the two alkene monomers. This allows the distinction to be made between the presence of radicals, anions or cations as the reactive species responsible for initiating the polymerization. Cations would give a polymer comprising styrene residues only; anions would give a polymer containing methyl methacrylate residues only. The three possibilities can be readily be distinguished by elemental analysis by combustion, which typically gives the carbon and hydrogen content to about 0.4%. Expected compositions of the polymers formed by these three methods of initiation are shown in Table 4.2 (Section 4.3.4).

By-products. The presence in the products of small quantities of compounds which would arise from combination of free radical intermediates can provide evidence for a free radical process. For example, the explosive reaction of methane with fluorine gives mainly hydrogen fluoride and a mixture of mono-, di-, tri- and tetrafluoromethanes, but small quantities of fluorinated ethanes, including C_2F_6, are also produced. These two-carbon products cannot be readily explained on the basis of possible molecular reactions (see reaction 6.16), but would arise naturally as combination products of the fluorinated methyl radicals produced in a radical chain reaction sequence (reaction 6.17).

$$\left[\begin{matrix} CH_3 & F \\ | & | \\ H & F \end{matrix} \xrightarrow{\quad} \begin{matrix} CH_3-F \\ H-F \end{matrix} \xrightarrow{F_2} CH_2F_2 \xrightarrow{F_2} CHF_3 \xrightarrow{F_2} CF_4 \right] \qquad (6.16)$$

$$\left. \begin{aligned} CH_4 + F^\bullet &\longrightarrow CH_3^\bullet + H-F \\ CH_3^\bullet + F-F &\longrightarrow CH_3-F + F^\bullet \\ CH_3F \rightarrow \dot{C}H_2F \rightarrow CH_2F_2 \rightarrow \dot{C}HF_2 \rightarrow CHF_3 \rightarrow \dot{C}F_3 \xrightarrow{\dot{C}F_3} CF_3-CF_3 \end{aligned} \right\} \quad (6.17)$$

6.4   Electron Spin Resonance (ESR)

ESR spectra of radicals show some similarities with the NMR spectra of spin-paired molecules. The ESR *g* value corresponds to the NMR chemical shift. Hyperfine splitting in ESR caused by electron–nuclear spin interactions have an NMR counterpart in the splitting of resonances by interactions between two nuclear spins.

With specialized equipment, it is possible to investigate the presence of particular radical species by UV or visible spectroscopy, but the most useful spectroscopic method for free radicals is **electron spin resonance** or **ESR**. This technique depends on the fact that a free electron has a magnetic moment. In a magnetic field, this will line up either with the field or against it. These two orientations have slightly different energies, so if radiation of an appropriate frequency is applied, there will be absorption of energy at a frequency corresponding to the energy difference. The energy gap between the two spin states will be proportional to the magnitude of the applied field (see Figure 6.2a), and equation (6.18) shows the relationship between the frequency of radiation v and the applied magnetic field H. β is the Bohr magneton, a parameter derived from the properties of the electron, h is Planck's constant, and g, the dimensionless g value, has a value of 2.0023 for the free electron. Magnetic fields in the region of 300 mT are used, corresponding to radiation in the microwave region of about 3 cm. At this wavelength, which depends on the size of the cavity of the spectrometer, it is easier to keep the wavelength constant and sweep the magnetic field to achieve the resonance condition at which absorption of radiation can take place. Because of the method of detection, the display is in the form of the derivative of the absorption, giving a spectrum shown in Figure 6.2(b). The g value corresponds to the field at which the derivative curve crosses the baseline. Values of g vary over a small range and are larger for organic radicals with, for example, oxygen substituents. The g value corresponds to the chemical shift in the analogous NMR technique, but is nothing like as useful, since it is difficult to predict the effect of a particular substituent. The g value, is, however, characteristic of a particular radical even if prepared in different ways, provided the solvent and temperature remain the same.

$$E = hv = g\beta H \qquad (6.18)$$

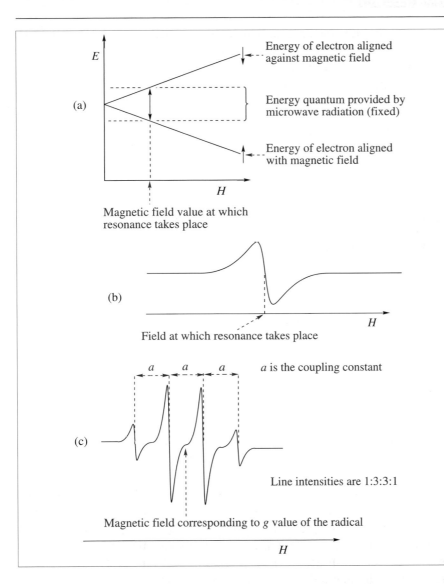

Figure 6.2 Electron spin resonance. (a) Dependence of electron energy on magnetic field. (b) ERS spectrum of a simple free radical. (c) Idealized ESR spectrum of the methyl radical

The presence of an ESR signal is conclusive evidence for the presence of a free radical in the system. The presence of an ESR signal in a reacting system is, however, not by itself conclusive evidence that the main body of the reaction is taking place by a free radical process, since a small amount of a side reaction could lead to a signal. Because of the small magnitude of the radical concentration often present, it is also possible to have a free radical reaction without producing a visible ESR signal.

If g values were the only information available from ESR, the technique would have limited use in showing the presence or absence of free radicals in a system. Fortunately, there is another feature. Many atomic nuclei have a nuclear magnetic moment, much smaller than that of the electron.

The most useful is the proton, which has a nuclear spin of ½, as do ^{19}F, ^{31}P and also ^{13}C (which is present in about 1% in natural abundance). ^{14}N has a spin of 1, and both ^{35}Cl and ^{37}Cl have spins of 3/2. ^{16}O has no nuclear spin. A list of common nuclei with nuclear spins is given in Table 6.1.

Table 6.1   Some common nuclei with nuclear spins

Atom	Nuclear spin	Natural abundance (%)
^1H	½	99.98
^{13}C	½	1.1
^{14}N	1	99.6
^{19}F	½	100
^{29}Si	½	4.7
^{31}P	½	100
^{35}Cl	3/2	75.5
^{37}Cl	3/2	24.5

If an atom with a nuclear spin of ½ is present close to the radical site, it will align with (↑) or against (↓) the magnetic field, augmenting or decreasing the magnetic field "seen" by the electron, so there will be two resonance positions and the spectrum will show a doublet of lines of equal intensity. Two identical atoms in the same environment will behave identically, giving three possible orientations and will therefore produce three lines. However, since the possible orientations are ↑↑, ↑↓, ↓↑ or ↓↓, with equal probabilities, the central line, corresponding to the two situations ↑↓ and ↓↑ where there is no net effect of the two nuclei, will have twice the intensity of the outside lines, leading to a triplet with relative intensities of the lines of 1:2:1, analogously to NMR. Likewise, three hydrogen atoms, for example in a methyl group, will give rise to a quartet with intensities 1:3:3:1, and in general n hydrogen atoms will give $n + 1$ lines. An idealized spectrum of the methyl radical is shown in Figure 6.2(c).

If more than one environment for the hydrogen is present, each type can be considered separately, so that in the ethyl radical $CH_3CH_2{}^{\bullet}$, each line of the triplet due to the methylene group is split into quartets by the three hydrogens of the methyl group, giving a spectrum of 12 lines in all.

Atoms such as ^{14}N and deuterium with a nuclear spin of 1 have three possible orientations of the nuclear magnetic moment, either with the field, at right angles to it, or opposed to the field. This gives rise to a 1:1:1 triplet, in contrast to the 1:2:1 triplet caused by two identical protons.

These splitting patterns often allow identification of a radical and to determine the position of attack on a molecule. For example, t-butoxyl radicals react with diethyl ether to give a product radical whose spectrum shows a quartet of doublets of triplets, corresponding to three different kinds of hydrogen atoms containing one, two and three atoms,

Worked Problem 6.1

Q The ESR spectrum of the benzene radical anion, $[C_6H_6]^{\cdot-}$, in which the unpaired electron is in a molecular orbital delocalized round the benzene ring, shows a septet, with a coupling constant of 0.375 mT. Comment on the spectrum, in relation to the quartet shown by the methyl radical, with $a(C–H) = 2.30$ mT.

A The quartet in methyl is due to three identical hydrogen atoms; n hydrogens give $n+1$ lines. Six identical hydrogens in benzene should give a septet, as observed. The coupling constant reflects spin density on the appropriate carbon atom. In methyl, the spin density is almost all on the central carbon atom. In the benzene radical anion it will be equally shared on all the carbon atoms, giving a spin density of $1/6$ on each carbon atom. The coupling constant should therefore be $23.0/6 = 0.383$ mT, in very good agreement with the observed value of 0.375 mT.

6.5.3 β-Hydrogen Coupling Constants

These have the same order of magnitude as the α-coupling constants. For rigid radicals, the value depends critically on the orientation of the C–H bond with the orbital containing the unpaired electron (see Figure 6.5). If the σ-orbital of the C–H bond is exactly aligned with the axis of the p orbital containing the unpaired electron (**34**, dihedral angle $\theta = 0°$), there will be maximum interaction, spin density will be transferred to the hydrogen and a large coupling, in the region of 5 mT, will be observed. If the dihedral angle is 90°, conformation **35**, there will be no interaction and no coupling. For intermediate angles, the coupling constant will depend on $\cos^2\theta$. If there is free rotation round the C–C bond, for example in the ethyl radical, there will be an average over all possible dihedral angles. Since the average value of $\cos^2\theta$ over all angles from 0° to 360° is 0.5, the expected value of the coupling constant would be half the maximum value. The β-C–H coupling constant for ethyl is 2.7 mT, in approximate agreement with the expected value of $0.5 \times 5 = 2.5$ mT.

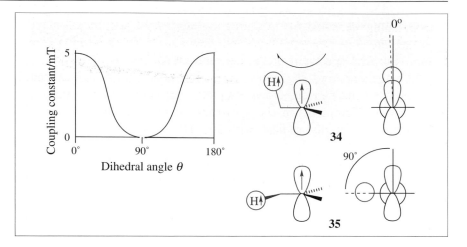

Figure 6.5 Dependence of β-hydrogen coupling constants in organic radicals on the dihedral angle θ

On a resonance picture, we can consider structures such as **36a** and **36b** as contributing to the total structure. Stabilization of this type is referred to as hyperconjugation. It must be stressed that structures such as **36b** only contribute to the structure to a minor extent, so there is no question of the hydrogen atom becoming free and leaving the radical.

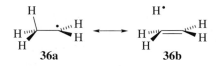

Hydrogen atoms more remote than the β-position have coupling constants which are typically at least an order of magnitude less than α and β coupling constants, and are usually unimportant.

Worked Problem 6.2

Q The ESR spectrum of the cyclopentyl radical is a doublet ($a = 2.1\,\text{mT}$) of quintets ($a = 3.5\,\text{mT}$). What does this show about the structure of the radical? [The ethyl radical shows a triplet ($a = 2.2\,\text{mT}$) of quartets ($a = 2.7\,\text{mT}$)].

A Saturated alkyl radicals are approximately planar at the radical centre, with most of the spin density on the central atom, so the doublet splitting, ascribed to the C–H at the radical centre, has a similar value for that of the triplet in ethyl due to the CH_2 group. Because of the ring, the four β-hydrogen atoms in cyclopentyl are not free to rotate. The dihedral angle between the β-hydrogen atoms and the unpaired electron is about 30°. $\cos^2 30° = 0.75$, so the quintet coupling constant due to the four β-hydrogen atoms should be approximately $5 \times 0.75 = 3.75$, in good agreement with the observed value, and larger than $a(CH_3)$ in ethyl, where the free rotation of the methyl groups leads to an average value of $\cos^2\theta = 0.5$.

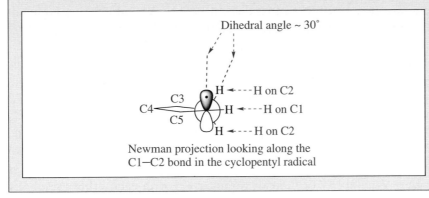

Newman projection looking along the
C1—C2 bond in the cyclopentyl radical

6.5.4 Stability and Persistence

Stability is a thermodynamic concept, and is defined for organic radicals as the difference between the bond dissociation energies of R–H and CH_3–H. Values of $D(R–H)$ for a number of radicals are given in Table 6.3. The stability increases in the order $CH_3^{\bullet} < CH_3CH_2^{\bullet} < (CH_3)_2CH^{\bullet} < (CH_3)_3C^{\bullet}$. This increase in stability is normally ascribed to hyperconjugation as described above, but the increasing number of alkyl carbon atoms round the free radical centre will also mean that these sp^3–sp^3 C–C bonds will be converted into the stronger sp^3–sp^2 bonds as the branched alkane is converted into the radical, which will decrease the bond dissociation energy and therefore increase the stability of the radical.

Table 6.3 Bond dissociation energies for hydrocarbons, $D(R–H)$[a]

R	$D(R–H)/\text{kJ}\,\text{mol}^{-1}$
Me$^{\bullet}$	439
MeCH$_2^{\bullet}$	423
Me$_2$CH$^{\bullet}$	412
Me$_3$C$^{\bullet}$	404
CH$_2$=CHCH$_2^{\bullet}$	362
PhCH$_2^{\bullet}$	370
Ph$_2$CH$^{\bullet}$	341
Ph$_3$C$^{\bullet}$	298
H$_2$C=CH$^{\bullet}$	465
Ph$^{\bullet}$	465
HC≡C$^{\bullet}$	556

[a]Secondary and tertiary C–H bonds in alkanes have strengths very similar to those of MeCH$_2^{\bullet}$, Me$_2$CH$^{\bullet}$ and Me$_3$C$^{\bullet}$, respectively

The bond from H to sp^2 carbon is stronger than that to sp^3 carbon so the vinyl (ethenyl), ethynyl and phenyl radicals are all less stable than the methyl radical. On the other hand, delocalized radicals such as allyl and benzyl are significantly more stable than methyl. The delocalization is demonstrated by ESR, so that delocalization and stabilization go hand-in-hand. Delocalization is increased if the radical can be spread over more than one ring, as in diphenylmethyl and triphenylmethyl.

Stability as defined in this way is a good guide to reactivity, since the trend in bond strengths to other elements broadly follows the strengths to carbon. Thus the more stabilized the radical, the less reactive it will be towards a particular reagent.

Most radicals react with each other very rapidly by combination and disproportionation to give non-radical products, and if radicals are not continuously introduced into the system, the concentration quickly drops to zero. A minority of radicals do not react with each other rapidly by combination or disproportionation, usually for steric reasons or because no bond of sufficient strength can be formed. Radicals of this type which can maintain their concentration at least for a few hours in solution in inert solvents in the absence of oxygen are said to be **persistent**. Examples of persistent radicals include triphenylmethyl (**37**). Here, the central carbon atom is so crowded that dimerization to hexaphenylethane is impossible. Attack by the central carbon radical centre of one radical on the *para* position of another can take place, giving rise to an alternative dimer **38** that exists in equilibrium with the triphenylmethyl radical (reaction 6.23). Nitroxides such as 2,2,6,6-tetramethylpiperidinoxyl (TEMPO, **39**) are persistent. The lone pair orbital on the nitrogen and the singly occupied orbital on the oxygen interact to form a π bond; since three electrons are involved, one has to go into the π* orbital (Figure 6.6, **39a**). As a result, the N–O bond has a bond order of 1 ½, and this strength would be reduced to 1 in any addition or abstraction reaction of the oxygen atom.

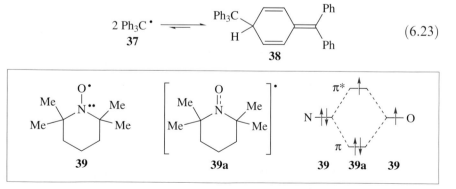

$$2\ Ph_3C^{\bullet} \rightleftharpoons 38 \tag{6.23}$$

Figure 6.6   The TEMPO radical

Radicals can be stable without being persistent; the benzyl radical **33** is stabilized by delocalization of the electron onto the *ortho* and *para* positions of the benzene ring, but it is not persistent since dimerization can readily take place at the unhindered CH_2 group.

6.6   Radical Chain Reactions

The high reactivity and fast recombination rate of most radicals ensures that however rapidly radicals are introduced into a system, the steady-state concentration of radicals remains low. Under these conditions, a radical will meet thousands of molecules before it is destroyed on meeting another radical, offering excellent opportunities for reaction. Because of the single electron involved, reaction of a radical with an electron paired molecule must produce another radical, which in turn can react with another molecule to produce another radical, and so on, until eventually the radical is destroyed on encounter with another radical. If the radical is reactive and the radical concentration is very low, many thousands of molecules may be converted for each radical introduced into the system. Reactions of this type are called **chain reactions**; almost all free radical reactions of synthetic or environmental importance are chain reactions. Because of their importance, a terminology for chain reactions has been developed. **Initiation** reactions are those in which a radical or radicals are introduced into the system (see Section 6.1). **Termination** reactions are those in which radicals are destroyed, usually combination and disproportionation (Section 6.2). **Propagation** reactions are the radical–molecule reactions in which all the useful chemistry takes place, in favourable cases at least 99%, leaving less than 1% of by-products from termination reactions. There are two important classes of propagation reactions: radical transfer and addition to a multiple bond (and its reverse, fragmentation).

Almost all useful radical reactions are chain reactions. The major products are determined by the propagation steps. Only minor amounts of by-products come from initiation or termination reactions.

6.6.1   Chain Reactions involving Radical Transfer

Radical transfer reactions involve abstraction of an atom or group B by a radical A from a molecule B–C (reaction 6.24). B is nearly always an atom; transfer of a group, which would correspond to substitution at a polyvalent atom, though important in nucleophilic and electrophilic reactions, is very uncommon in radical reactions. The atom transferred is almost always a hydrogen or a halogen atom.

$$A^{\bullet} + B–C \longrightarrow A–B + C^{\bullet} \qquad (6.24)$$
B is normally a hydrogen or a halogen atom

Reactions such as (6.25) which involve transfer of an oxygen atom are important in atmospheric chemistry. They are thought to involve addition of the hydroxyl radical to form an intermediate **40**, which then loses a hydrogen atom.

$$H-O^{\bullet} + {}^{-}C\equiv O^{+} \quad (\longrightarrow) \quad H^{\bullet} + O=C=O$$

$$H-O-\overset{\bullet}{C}=O$$
$$\mathbf{40}$$

(6.25)

Reactions such as (6.26) involve transfer of a hydrogen atom from a radical to a multiply bonded atom, and can be seen as hopping of a hydrogen atom between a C=O and an O=O double bond.

$$^{\bullet}O-CH_2-H + O=O \longrightarrow O=CH_2 + H-O-O^{\bullet}$$

(6.26)

Radical transfer reactions involved in chain reactions are almost always exothermic. Endothermic reactions would be faster in the reverse direction. This principle can sometimes be used to generate a more stable radical from a less stable radical. Thus the non-stabilized t-butoxyl radical **41** reacts with toluene to give t-butanol and the benzyl radical **42** (reaction 6.27).

$$Me_3C-O^{\bullet} + H-CH_2-Ph \longrightarrow Me_3C-O-H + {}^{\bullet}CH_2-Ph$$
$$\mathbf{41} \hspace{7cm} \mathbf{42}$$

(6.27)

Halogenation

Halogenations of alkanes provide good examples of chain reactions where all the chain steps involve radical transfer. Initiation is by photolysis of the halogen (reaction 6.28; X = Cl or Br) or, for fluorine, by molecule-induced homolysis (reaction 6.10). There are two propagation steps. The first (6.29) involves abstraction of a hydrogen atom from the alkane (illustrated here by methane). The alternative reaction to give the halomethane and a hydrogen atom does *not* take place: this would involve attack at a multivalent centre which is very uncommon for organic compounds without electropositive substituents, and the reaction would be $90 \, kJ \, mol^{-1}$ endothermic for $Cl^{\bullet}$, which would lead to an activation energy of at least that amount. The second propagation step (6.30) involves abstraction of a halogen atom from a halogen molecule by the methyl radical. Termination involves combination reactions of $X^{\bullet}$ and $CH_3^{\bullet}$ to give X–X, CH_3X and CH_3CH_3. The feasibility of the chain reaction depends on both propagation steps being fast. This is partly controlled by the thermochemistry; heats of reaction for both steps of the chain reaction are shown in Table 6.4 for all four halogens.

$$X-X \xrightarrow{h\nu} X^{\bullet} + X^{\bullet}$$
$$X = Cl \text{ or } Br$$

(6.28)

$$CH_3-H + X^{\bullet} \longrightarrow CH_3^{\bullet} + H-X$$

(6.29)

$$CH_3^{\bullet} + X-X \longrightarrow CH_3-X + X^{\bullet}$$

(6.30)

Table 6.4 Heats of reaction for halogenation of methane (kJ mol^{-1})

	Halogen F	Cl	Br	I
ΔH for reaction (6.29); hydrogen abstraction by X$^{\bullet}$	−133	+3	+69	+136
ΔH for reaction (6.30); reaction of CH$_3{}^{\bullet}$ with X–X	−298	−103	−99	−82
ΔH overall	−431	−100	−30	+54

For fluorine, both steps are highly exothermic. Each step is fast; the reaction is explosive. For chlorine, the first reaction is approximately thermoneutral, but the extra activation energy barrier is not high, and the first step is fast. The second step is substantially exothermic and fast; the overall exothermicity ensures that there is no tendency for the reaction to reverse. For bromine, the first step is substantially endothermic, and the activation energy must be greater than this. The reaction is therefore much slower than chlorination, even though the second step is highly exothermic, and will be expected to be fast. The overall exothermicity of the whole process will ensure that the reaction goes to completion. For iodine, the endothermicity of the first step makes it too slow for reaction to proceed. Additionally, the exothermicity of the second step is not sufficient to balance the endothermicity of the first step, so the reaction is endothermic overall. In fact, the reaction can be operated in reverse: hydrogen iodide will reduce iodoalkanes to alkanes with iodine as the other product.

For hydrocarbons with more than one type of C–H bond, halogenation will give mixtures of products. The position of attack is determined by the abstraction step, and the heat of reaction (and also the activation energy) will reflect the strength of the bond broken. Table 6.3 shows that bond strengths fall from primary > secondary > tertiary > allylic or benzylic carbon atoms, and we may therefore expect reactivity to rise in the same order, as the bond becomes easier to break. The situation was discussed in Chapter 3. For the very exothermic reactions of fluorine, there will be an early transition state (Figure 3.1c), which will be very close to the starting materials. There will be little development of radical character on the alkyl group, so there will be little differentiation because of the different stabilities of the incipient alkyl radicals. Thus attack will be essentially random (see Table 6.5). On the other hand, attack by bromine atoms will have a late transition state with a structure near that of the products, so that any radical stabilization will be almost fully developed. Bromination will therefore be highly selective, and the different reactivities of different positions enable useful preparative chemistry to be carried out.

Table 6.5 Relative selectivity in hydrogen abstraction from alkanes by halogen atoms at 300 K[a]

Halogen	$-CH_3$	$\diagdown CH_2 \diagup$	$\diagdown -CH \diagup$
F	1	1.2	1.4
Cl	1	4.4	6.7
Br	1	80	1600
I	1	1850	210,000

[a] Relative selectivities are for a single hydrogen of the type indicated.

Allylic positions are even more reactive, so that, for example, although cyclohexene has three types of hydrogen atom, bromination takes place almost exclusively at the 3-position via the allylic radical **43** (reaction 6.31).

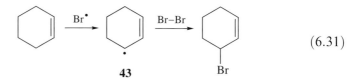

$$(6.31)$$

43   Br

Organotin Hydride Reductions

Tin forms strong covalent bonds to halogens but only a weak bond to hydrogen. This feature can be exploited in a radical chain reaction that results in reduction of haloalkanes to alkanes by organotin hydrides; the chain steps are shown in reactions (6.32) and (6.33). In the initiation step, a radical from homolysis of an azo compound (*e.g.* reaction 6.2) abstracts the weakly bound hydrogen atom from the tin centre of **44**. The organotin radical **45** preferentially abstracts a halogen atom from the haloalkane to give the alkyl radical, and does not react with the C–H bonds because the Sn–H bond is so weak. In the second step, the alkyl radical preferentially abstracts a hydrogen atom from the Sn–H bond, the weakest bond in the two molecules. Thus haloalkanes can be reduced to alkanes **46** in good yields, and the reaction can be carried out in the presence of groups such as the carbonyl group, which is readily reduced by polar reagents such as lithium aluminium hydride, $Li^+[AlH_4]^-$.

$$Bu_3Sn^{\bullet} + Br-R \longrightarrow Bu_3Sn-Br + R^{\bullet} \qquad (6.32)$$
$$\phantom{Bu_3Sn^{\bullet} + }\mathbf{45}$$

$$R^{\bullet} + H-SnBu_3 \longrightarrow R-H + {}^{\bullet}SnBu_3 \qquad (6.33)$$
$$\phantom{R^{\bullet} + H-}\mathbf{44}\mathbf{46}\phantom{-H + {}^{\bullet}Sn}\mathbf{45}$$

6.6.2 Chain Reactions involving Addition to Multiple Bonds (and Fragmentation)

The second important type of propagation reaction is addition to multiple bonds; addition to C=C is particularly important. In reaction (6.34), R can be an atom or a group centred on carbon or any element which forms a bond stronger than the π bond which is broken in the reaction (about $250 \, kJ \, mol^{-1}$). If the alkene is unsymmetrical, addition can in principle take place at either end of the double bond. Addition normally takes place at the end of the double bond which will generate the more stable free radical. Thus for addition of a halogen atom to propene, attack at the CH_2 position will give the secondary radical **47** (reaction 6.35) rather than attack at the central carbon atom which would give the less stable primary radical **48** (reaction 6.36).

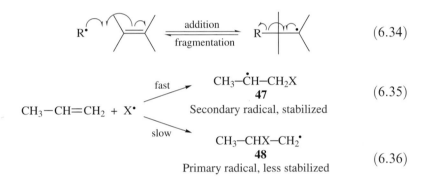

$$(6.34)$$

$$CH_3-\overset{\bullet}{C}H-CH_2X$$
47
Secondary radical, stabilized

$$(6.35)$$

$$CH_3-CH=CH_2 + X^\bullet$$

$$CH_3-CHX-CH_2^\bullet$$
48
Primary radical, less stabilized

$$(6.36)$$

The most favourable approach for attack by a radical appears to be along the line on which the new bond is being formed (Figure 6.7). For the intramolecular addition of the hex-5-enyl radical **49**, this results in the formation of the less stable primary cyclopentylmethyl radical **50** by reaction (6.37) rather than the more stable secondary cyclohexyl radical **51** by reaction (6.38).

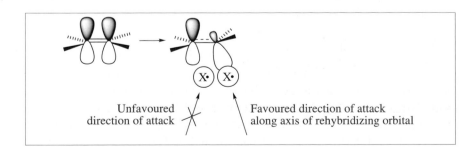

Unfavoured direction of attack

Favoured direction of attack along axis of rehybridizing orbital

Figure 6.7 Direction of approach of a radical to a double bond

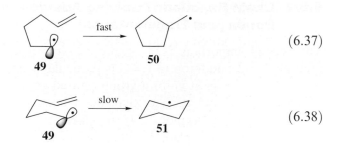

$$(6.37)$$

$$(6.38)$$

Fragmentation

Addition reactions differ from radical transfer reactions in one very important respect: two reactants react to give one product radical. This decrease in number of particles means that entropy is decreased during addition. The reverse reaction, fragmentation, involves an increase in the number of particles and therefore in the entropy. Since the free energy of reaction, ΔG, and therefore the position of equilibrium is governed by equation (6.39), however exothermic the addition reaction is, at a sufficiently high temperature the $T\Delta S$ term will outweigh ΔH and fragmentation will take place. t-Butoxyl radicals **52** fragment to methyl radicals and acetone at $140\,^{\circ}\mathrm{C}$ (reaction 6.40), and fragmentation of alkyl radicals will be considered in the next section on polymerization.

Negative ΔG values correspond to equilibria that lie to the right.

$$\Delta G = \Delta H - T\Delta S = -RT\ln K \qquad (6.39)$$

$$(6.40)$$

Polymerization

The radical product of addition to a double bond may itself be able to add to a new alkene molecule to give a longer-chain radical, and this process may be repeated in a chain reaction consisting only of radical additions to give a long-chain polymer radical. The process to give the **polymer** is completed by termination steps involving combination or disproportionation of two polymer radicals. This chain-reaction free radical polymerization process is exemplified in reactions (6.41)–(6.47) for the polymerization of styrene initiated by azobisisobutyronitrile (**53**). The 2-cyanoprop-2-yl radical **54** adds to the CH_2 group of a styrene molecule to give the stabilized radical **55** (reaction 6.42). The adduct radical **55** in turn adds to the CH_2 group of another styrene molecule to give the radical **56** (reaction 6.43), and by successive additions (6.44 and 6.45) a long-chain polymer radical **57** is produced, as a straight chain with phenyl substituents on alternate backbone carbon atoms. Termination takes place by combination (reaction 6.46) to give a dimeric polymer **58**, or by disproportionation where one polymer radical **57** abstracts a hydrogen atom from a second radical to give an alkane **59** and an alkene **60** polymer molecule (reaction 6.47), both with shorter chain lengths than polymer molecules produced by combination. Useful polymers tend to have chain lengths of at least 1000. Most alkenes that can be readily polymerized have a terminal CH_2 or CF_2 group.

Initiation

$$Me_2C(CN)-N=N-C(CN)Me_2 \longrightarrow 2\ Me_2\overset{\bullet}{C}-CN + N_2 \tag{6.41}$$

53   **54**

$$In^{\bullet}$$

$$In^{\bullet} + CH_2=CH-Ph \longrightarrow In-CH_2-\overset{\bullet}{C}H-Ph \tag{6.42}$$

55

Propagation

$$In-CH_2-\overset{\bullet}{C}H-Ph + CH_2=CH-Ph \longrightarrow In-CH_2-\overset{\overset{\displaystyle Ph}{|}}{C}H-CH_2-\overset{\bullet}{C}H-Ph \tag{6.43}$$

56

$$In-CH_2-\overset{\overset{\displaystyle Ph}{|}}{C}H-CH_2-\overset{\bullet}{C}H-Ph + CH_2=CH-Ph \longrightarrow$$

56

$$In\left(CH_2-\overset{\overset{\displaystyle Ph}{|}}{C}H\right)_2 CH_2-\overset{\bullet}{C}H-Ph \tag{6.44}$$

$$\overset{\longrightarrow}{\Longrightarrow} In\left(CH_2-\overset{\overset{\displaystyle Ph}{|}}{C}H\right)_n CH_2-\overset{\bullet}{C}H-Ph \tag{6.45}$$

$$R_{poly}-CH_2-\overset{\bullet}{C}H-Ph$$

57

Termination

$$2\ R_{poly}-CH_2-\overset{\bullet}{C}H-Ph \longrightarrow R_{poly}-CH_2-\overset{\overset{\displaystyle Ph}{|}}{C}H-\overset{\overset{\displaystyle Ph}{|}}{C}H-CH_2-R_{poly} \tag{6.46}$$

57   **58**

Combination

$$R_{poly}-CH_2-CH_2-Ph + R_{poly}-CH_2=CH-Ph \tag{6.47}$$

59 Disproportionation **60**

The average chain length of polymers can be reduced by adding small quantities of **chain transfer agents** such as thiols. These react with the growing polymer chain by hydrogen transfer (reaction 6.48), to form a polymer molecule **59** and a thiyl radical, which can re-initiate further polymerization. Thus each original initiator radical results in several shorter chain polymer molecules rather than one long one. This enables the chain lengths of polymers to be controlled.

Chain transfer

$$R_{poly}-CH_2-\overset{\bullet}{C}H-Ph \ + \ RSH \ \longrightarrow \ R_{poly}-CH_2-CH_2-Ph \ + \ RS^{\bullet} \qquad (6.48)$$
$$\quad\quad\quad\quad\quad\quad \textbf{57} \quad\quad\quad\quad\quad\quad\quad\quad\quad\quad\quad\quad\quad\quad\quad \textbf{59}$$

Polymerizations are carried out at relatively low temperatures. Since the propagation steps (6.44) and (6.45) involve two starting molecules giving one product molecule, entropy is lost when the reaction goes in the forward direction. Equation (6.39) shows that as the temperature rises, so does the $T\Delta S$ contribution to the free energy of reaction, so that at a sufficiently high temperature the reverse of steps (6.44) and (6.45) will be faster than the forward reactions. Above this **ceiling temperature**, polymer molecules will degrade to monomer molecules by fragmentation. Ceiling temperatures are different for different polymers; for polystyrene [poly(phenylethene)] the temperature is $310\,^{\circ}C$.

Addition of HBr to Alkenes

In the presence of a radical initiator, alkenes react with reactive molecules such as hydrogen bromide to give simple 1:1 adducts rather than a polymer. The initiator radical reacts rapidly with an HBr molecule to give a bromine atom (6.49), which starts the chain reaction. In the first propagation step, the bromine atom adds to the alkene **61** to give the adduct radical **62** (reaction 6.50). Since **62** abstracts a hydrogen atom from HBr by reaction (6.51) more rapidly than it would add to the alkene to form a polymer radical as in (6.43), the chain continues with reactions (6.50) and (6.51) as the propagating steps, and the product is the primary bromo compound **63**. This **anti-Markovnikov** addition is in the reverse direction to the polar addition discussed in Chapter 5. Since the radical chain reaction is faster than the polar reaction, the anti-Markovnikov product dominates if radicals are present. If the Markovnikov product is required, the reaction must be carried out in the dark, in the absence of free radical initiators, and preferably with a radical inhibitor present.

$$In^{\bullet} \ + \ H-Br \ \longrightarrow \ In-H \ + \ {}^{\bullet}Br \qquad\qquad (6.49)$$

$$R-CH=CH_2 \ + \ {}^{\bullet}Br \ \longrightarrow \ R-\overset{\bullet}{C}H-CH_2-Br \qquad\qquad (6.50)$$
$$\quad\quad \textbf{61} \quad\quad\quad\quad\quad\quad\quad\quad\quad \textbf{62}$$

$$R-\overset{\bullet}{C}H-CH_2-Br \ + \ H-Br \ \longrightarrow \ R-CH_2-CH_2-Br \ + \ {}^{\bullet}Br \qquad (6.51)$$
$$\quad \textbf{62} \quad\quad\quad\quad\quad\quad\quad\quad\quad\quad\quad \textbf{63}$$

Propagation

Worked Problem 6.3

Q Photolysis of a mixture of CCl_3Br and the alkene $PhCH=CH_2$ gives $PhCHBr-CH_2CCl_3$ as the major product. Write a plausible chain reaction scheme to account for this product.

A

$$Br-CCl_3 \xrightarrow{h\nu} Br^{\bullet} + {}^{\bullet}CCl_3 \qquad \text{Initiation}$$

$$Ph-CH=CH_2 + {}^{\bullet}CCl_3 \longrightarrow Ph-\overset{\bullet}{C}H-CH_2-CCl_3$$
addition at CH_2 gives
the more stable radical

Propagation

$$Ph-\overset{\bullet}{C}H-CH_2-CCl_3 + Br-CCl_3 \longrightarrow Ph-CHBr-CH_2-CCl_3 + {}^{\bullet}CCl_3$$
univalent bromine
atom is transferred

What happens to the bromine atom produced in the initiation step? It will react with the alkene as shown below, and the adduct radical will react with CCl_3Br to give a ${}^{\bullet}CCl_3$ radical, which will continue the chain as shown above. One molecule of the dibromide will be produced as a by-product, along with termination products. Provided that the chain is long, these will be insignificant.

$$Ph-CH=CH_2 + {}^{\bullet}Br \longrightarrow Ph-\overset{\bullet}{C}H-CH_2-Br$$

$$Ph-\overset{\bullet}{C}H-CH_2-Br + Br-CCl_3 \longrightarrow Ph-CHBr-CH_2-Br + {}^{\bullet}CCl_3$$

6.7 Atmospheric Reactions

Almost all the chemistry that takes place in the atmosphere is free radical in nature. An important area is the oxidative degradation of organic matter, ultimately to carbon dioxide, with formaldehyde (methanal) and carbon monoxide as key intermediates. The hydroxyl radical is the main daytime initiating radical, formed principally by the photolysis of NO_2 to give an oxygen atom which reacts with water to give two hydroxyl radicals (reaction 6.52). A simplified scheme for oxidation of methane to formaldehyde in an unpolluted atmosphere is shown in reactions (6.53)–(6.57). This sequence of five elementary steps involves two hydrogen transfer reactions, an addition to a double bond, a disproportionation reaction and a photolysis, all of which are analogous to reactions already considered in this chapter. The overall result (equation 6.58) is the oxidation of one molecule of methane by a molecule of oxygen to formaldehyde and a water molecule. One photon is required, and the $^\bullet$OH, $CH_3O^\bullet$ and $CH_3OO^\bullet$ radicals which are formed during the reaction sequence are destroyed in other steps. The disproportionation reaction (6.55) is thought to be the way peroxyl radicals are destroyed: simple combination would lead to a molecule containing a chain of four oxygen atoms which would be unstable and immediately break in the centre to re-form the peroxyl radicals.

$$NO_2 \xrightarrow[\lambda < 400\ nm]{h\nu} NO + O \xrightarrow{H_2O} 2\ {}^\bullet OH \qquad (6.52)$$

$$CH_4 + {}^\bullet OH \longrightarrow CH_3^\bullet + H_2O \qquad (6.53)$$

$$CH_3^\bullet + O{=}O \longrightarrow CH_3{-}O{-}O^\bullet \qquad (6.54)$$

$$CH_3{-}O{-}O^\bullet + {}^\bullet O{-}O{-}H \longrightarrow CH_3{-}O{-}O{-}H + O{=}O \qquad (6.55)$$

$$CH_3{-}O{-}O{-}H \xrightarrow{h\nu} CH_3{-}O^\bullet + {}^\bullet O{-}H \qquad (6.56)$$

$$H{-}CH_2{-}O^\bullet + O{=}O \longrightarrow CH_2{=}O + {}^\bullet O{-}O{-}H \qquad (6.57)$$

$$CH_4 + O_2 \xrightarrow{h\nu} H_2CO + H_2O \qquad (6.58)$$

Chemistry in the atmosphere is almost entirely free radical reactions.

Formaldehyde and other aldehydes and ketones formed during the atmospheric oxidation of more complex organic compounds are photolysed with a loss of a hydrogen atom or an alkyl radical. Photolysis of ketones (6.59) provides one way in which C–C bonds break; another is the fragmentation of alkoxyl radicals mentioned earlier (reaction 6.40). Eventually, all organic compounds end up as carbon dioxide and water.

$$CH_3{-}CO{-}CH_3 \xrightarrow{h\nu} CH_3^\bullet + {}^\bullet CO{-}CH_3 \longrightarrow CO + {}^\bullet CH_3 \qquad (6.59)$$

6.8   Non-chain Radical Reactions

Most useful free radical reactions are chain reactions and for photochemical reactions a large number of starting molecules are converted into products for every photon absorbed. A few photochemical processes are non-chain reactions, in which only one or two initial molecules are converted into products for each photon absorbed.

6.8.1 Nitrosation of Cyclohexane

The reaction of cyclohexane with nitrosyl chloride (NOCl) is a commercially important photochemical reaction, used in the first stage of nylon-6 manufacture. The UV light is absorbed by the NOCl (cyclohexane does not absorb UV light with $\lambda > 200$ nm) to give NO and Cl$^\bullet$ (reaction 6.60). The NO, though a radical, is unreactive, but the chlorine readily abstracts a hydrogen atom from cyclohexane (**64**) to give the cyclohexyl radical **65** (reaction 6.61). The cyclohexyl radical undergoes a combination reaction (6.62) with the NO, to give nitrosocyclohexane (**66**). One photon is required to produce one molecule of product; thus it is not a chain reaction. The remaining steps in the process to give the oxime **67**, the lactam **68** and eventually nylon-6 (**69**) are ionic. They are shown as reaction (6.63), but are not considered here.

$$Cl-N=O \xrightarrow{\;h\nu\;} Cl^\bullet + \,^\bullet N=O \qquad (6.60)$$

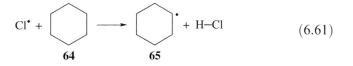

(6.61)

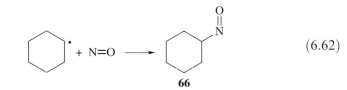

(6.62)

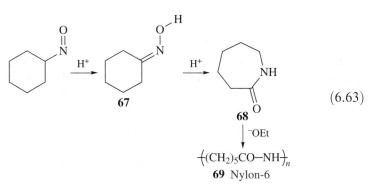

(6.63)

6.8.2 Intramolecular Hydrogen Abstraction: the Barton Reaction

Most organic compounds have more than one type of C–H bond, and hydrogen abstraction by radicals usually proceeds in an unselective manner, giving complex mixtures of products. An exception is when the abstraction is intramolecular, when for example a long-chain alkoxyl

Derek H. R. Barton shared the Nobel Prize for Chemistry in 1969 for his work on conformations of organic molecules.

radical **70**, produced in reaction (6.64), abstracts a hydrogen from its own chain to give **71**, as in reaction (6.65). Here the reaction takes place almost exclusively at the δ-position. The radical **71** combines with N=O to give the nitroso compound **72**. This is readily isomerized by acid to the oxime **73**, which can be hydrolysed to the ketone **74** (reaction 6.66). This allows the remote functionalization of a C–H group, and has been particularly useful in steroid chemistry. Yields are ~30%.

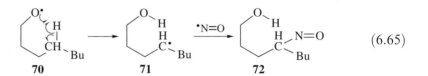

$$\text{C}_8\text{H}_{17}\text{—OH} \xrightarrow{\text{Cl—N=O}} \text{C}_8\text{H}_{17}\text{—O—N=O} \xrightarrow{h\nu} \underset{\textbf{70}}{\text{C}_8\text{H}_{17}\text{—O}^\bullet} + {}^\bullet\text{N=O} \qquad (6.64)$$

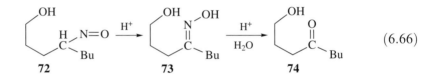

(6.65)

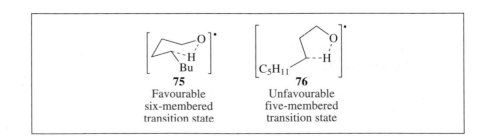

(6.66)

This reaction is not a chain reaction. One photon is required for each molecule of nitrite converted. The nitric oxide produced in the photolytic step is too unreactive to start a chain reaction, and simply combines with the rearranged radical.

The preferential abstraction of a δ-hydrogen atom corresponds to a six-membered transition state, which can adopt the unstrained cyclohexane chair-type conformation **75**. The alternative γ-hydrogen abstraction, which would require a five-membered transition state **76**, is not usually observed. By analogy with cyclopentane, this would be approximately planar, involving significant angle strain and eclipsing interactions which would raise the energy.

A few non-chain reactions are significant. They usually involve radical intermediates such as NO, which can combine with other radical intermediates, but are too stable to react with molecules in propagation steps.

75
Favourable
six-membered
transition state

76
Unfavourable
five-membered
transition state

Summary of Key Points

1. Most radical reactions are chain reactions. Initiation is usually by thermolysis of a weak bond or photolysis of a molecule with a suitable chromophore to absorb the light.

2. The most important reactions of radicals are atom abstraction and addition to multiple bonds. Atom abstraction is almost always hydrogen or halogen.

3. Fragmentation, the reverse of addition, becomes more important at higher temperatures.

4. Characteristics of free radical reactions include acceleration by free radical initiators or light, inhibition by compounds such as phenols, kinetics including half-powers of concentrations, and the production of small quantities of by-products resulting from radical combination.

5. Electron spin resonance is often helpful in deciding whether or not a reaction is free radical in character, and may identify intermediates.

6. Using these principles you should be able to devise possible reaction steps for a reaction that may involve free radical intermediates, and to devise tests to determine whether or not the reaction is free radical in nature.

Further Reading

D. C. Nonhebel and J.C. Walton, *Free-radical Chemistry*, Cambridge University Press, Cambridge, 1974.
R. P. Wayne, *Chemistry of Atmospheres*, Oxford University Press, Oxford, 2000.

Problems

6.1. When dibenzoyl peroxide **A** is heated in toluene (methylbenzene), the products include benzoic acid, benzene, carbon dioxide and 1,2-diphenylethane. Account for the formation of these products and comment on the reaction steps you propose.

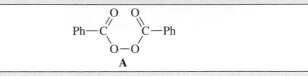

A

6.2. Methyl groups stabilize carbocations whereas nitro groups, $-NO_2$, stabilize carbanions. Provide an explanation, using the idea of delocalization, why both these substituents stabilize a radical centre conjugated with them.

6.3. The bond dissociation energies in Table 6.3 show that there is no significant stabilization of the phenyl radical compared with the vinyl (ethenyl) radical, $CH_2=CH^\bullet$, in spite of the fact that two structures **B1** and **B2** can be drawn for the phenyl radical. Explain this.

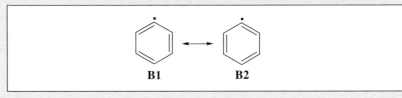

B1   B2

6.4. Dibenzylmercury in octane solution decomposes thermally to give a quantitative yield of mercury, with 1,2-diphenylethane as the major organic product. However, traces of toluene and 1,2,3-triphenylpropane are also produced. Suggest a mechanism. What other experiments could you carry out which might provide support for your mechanism?

6.5. When 2-methylprop-1-ene (isobutene, $Me_2C=CH_2$) is heated with thiophenol (benzenethiol, PhSH) in the presence of a small amount of di-t-butyl peroxide, the major product is Ph–S–CH_2CHMe_2. Suggest a mechanism for this reaction.

6.6. The ESR spectrum of the cyclohexyl radical consists of a doublet $(a = 2.1\,\text{mT})$ of triplets $(a = 3.9\,\text{mT})$ of triplets $(a = 0.5\,\text{mT})$. What light does this show on the structure of the radical?

6.7. When cyclohexyl nitrite **C** is photolysed, there is no intramolecular hydrogen transfer to give the nitroso compound **D** of the type found in reaction (6.44), whereas cyclooctyl nitrite **E** gives some rearranged product **F**. Why is this? [*Hint*: consider possible conformations].

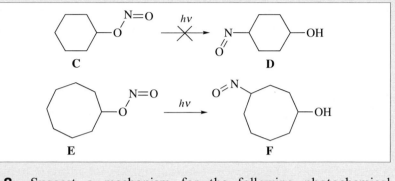

6.8. Suggest a mechanism for the following photochemical reaction [*Hint*: consider reactions (6.6) and (6.65)]:

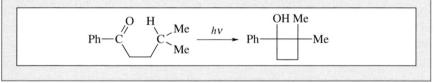

7
Molecular Reactions: Cyclic Transition States

The great majority of reactions involve intermediates, as illustrated in the previous three chapters. A smaller number of uni- or bimolecular reactions appear to proceed from reagents to products without detectable intermediates. These reactions take place in the gas phase or in solution in non-polar solvents. The absence of substantial solvent effects (Chapter 3) indicates little build up of charge in the transition state.

Butadiene dimerizes to 4-vinylcyclohexene (4-ethenylcyclohexene, **1**) (reaction 7.1). The absence of intermediates suggests a cyclic movement of three electron pairs, (which could equally well have been written in the opposite direction, or as single electron movements). The transition state would involve partial bond-making and breaking in the six-membered transition state as shown. Reactions involving such cyclic transition states are known as **pericyclic reactions**. However, ethene does not dimerize to cyclobutane (reaction 7.2) under thermal conditions, even though a cyclic movement of two pairs of electrons could have been invoked.

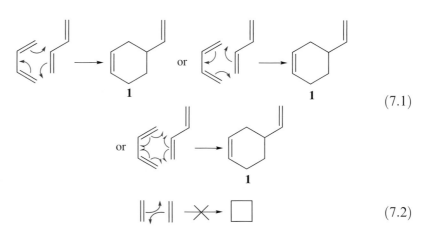

$$(7.1)$$

$$(7.2)$$

This chapter explores the reasons why some molecular reactions take place whereas others do not, and introduces the concepts of frontier orbitals and transition state aromaticity.

Aims

Having worked through this chapter, you should:

- Understand the concept of frontier orbitals, HOMO and LUMO, and be able to determine the HOMOs and LUMOs of organic molecules
- Understand the key importance of HOMO–LUMO interactions
- Understand the importance of Hückel and Möbius cyclic transition states
- Be able to apply these ideas to predict whether particular molecular reactions will or will not occur, and the stereochemical outcomes of such reactions

7.1 Frontier Orbitals

When two atoms approach, the atomic orbitals from the individual atoms begin to overlap, giving rise to molecular orbitals. The overlap is either in phase, giving rise to a lowering of energy resulting in bonding if the orbital is occupied by electrons, or out of phase, giving rise to an increase of energy and antibonding. This is illustrated in Figure 7.1 for two hydrogen atoms, each of which brings one 1s electron, both of which can be accommodated in the resulting bonding sigma (σ) orbital.

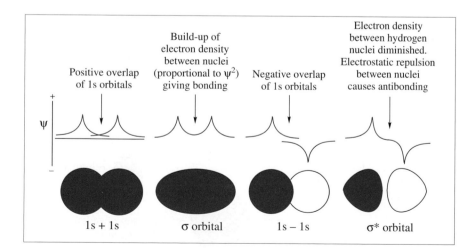

Figure 7.1 Overlap of the atomic orbitals of hydrogen. Positive phase for wave functions (orbitals) shown red, negative as white

Overlap of orbitals applies to molecules as well as atoms. In Figure 7.2 we see the approach of two hydrogen molecules end on. As the molecules approach, the σ orbitals interact to give two molecular orbitals, one with lower and one with higher energy. Since both original orbitals are occupied, with a total of four electrons, both the new molecular orbitals will be occupied. To a first approximation, the extra bonding and antibonding will cancel out, although because of electron repulsion the antibonding effect will be somewhat larger than the bonding effect.

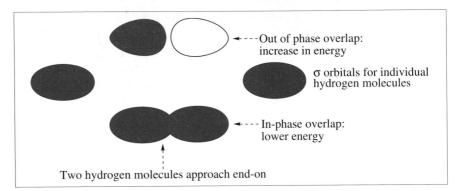

Figure 7.2 Overlap of the molecular orbitals of two hydrogen molecules

Worked Problem 7.1

Q What will happen to the σ orbital energies if two hydrogen molecules approach side to side?

A As the molecules approach, their σ orbitals will overlap. Two orbitals will result: the one with the phases the same will have a lower energy, whereas the one with different phases will have a higher energy. Since four electrons are involved, both new orbitals will be filled, so there will be no net increase in bonding.

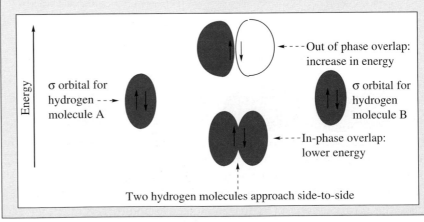

In most molecules, all the electrons are paired: orbitals are either filled with two electrons or empty. If two molecules approach each other, all the orbitals in one molecule will interact with all those in the other. For complex molecules with a large number of orbitals, this gives rise to a very large number of interactions. Fortunately, it turns out that only one or two are important.

The interactions are of three types: occupied–occupied, occupied–unoccupied and unoccupied–unoccupied (Figure 7.3). For occupied–occupied interactions, because four electrons are involved, both the resulting orbitals are filled. One goes down in energy, the other goes up, so there is no significant change in energy (Figure 7.3a). For unoccupied–unoccupied interactions there will be no energy consequences since neither of the new orbitals is occupied (Figure 7.3b). Only for occupied–unoccupied interactions is there a significant energy difference. Because only the more bonding new orbital is occupied, there will be a significant reduction in energy (Figure 7.3c).

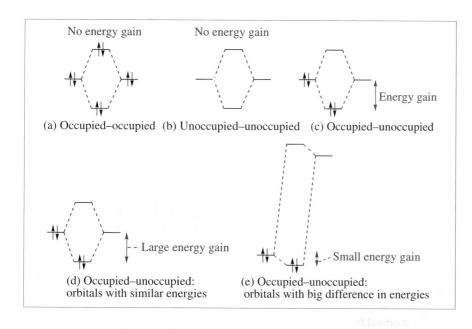

Figure 7.3 Energy consequences of the interaction between orbitals

Thus we should focus on occupied–unoccupied interactions. Interactions between orbitals are greater if the individual energies are similar, and less if they are significantly different (Figure 7.3d and 7.3e). *Thus the largest (favourable) energy interactions occur when there is the least energy difference between the interacting orbitals.* These correspond to interactions between the **highest occupied molecular orbital (HOMO)** of one molecule and the **lowest unoccupied molecular orbital (LUMO)** of the other (Figure 7.4). If the two reacting molecules have similar energy levels, the HOMO–LUMO interactions in each direction will both contribute significantly (Figure 7.4a), whereas if the levels are different in energy, the dominant interaction will be the one where there is the lesser difference in energies (Figure 7.4b).

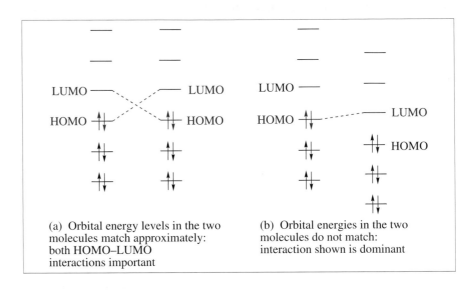

(a) Orbital energy levels in the two molecules match approximately: both HOMO–LUMO interactions important

(b) Orbital energies in the two molecules do not match: interaction shown is dominant

Figure 7.4 Frontier orbitals

It turns out that we can usually neglect all other orbital interactions between molecules and focus our attention only on these **frontier orbitals** (HOMOs and LUMOs) when considering interactions between molecules, and in particular pericyclic reactions. This insight gained the relatively unknown Japanese chemist Kenichi Fukui a share in the Nobel Prize for Chemistry in 1981.

How do we determine the frontier orbitals? We only need to consider the bonds that are actually reacting. For example, in reactions involving addition to ethene, we only need to consider the C–C π bond. The π orbital is the HOMO and the π^* orbital is the LUMO. For buta-1,3-diene with four orbitals in the π system, the HOMO and LUMO are shown in Figure 7.5b; note that there is a smaller HOMO–LUMO gap than for ethene (Figure 7.5a). In general, the antibonding level is approximately as far above the non-interacting level (separate atomic orbitals) as the bonding orbital is below this level.

In reaction (7.1), one of the butadiene molecules is reacting across its ends, so the HOMO and LUMO are for the whole conjugated system, but the other molecule is only reacting across one of the double bonds. The second double bond is unchanged in the reaction, so plays no part in

HOMO-LUMO interactions determine whether or not a particular thermal pericyclic reaction will take place.

the primary electronic interactions and can be regarded as a substituent of a reacting *ethene* molecule.

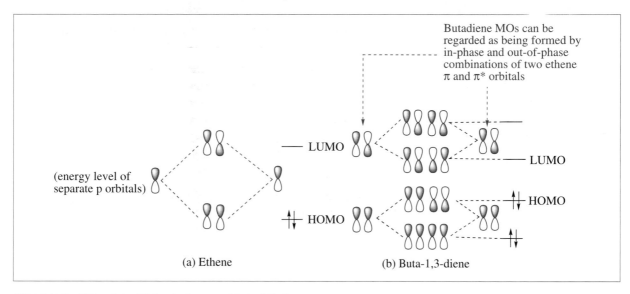

Figure 7.5 Molecular orbitals of ethene and buta-1,3-diene

In the next section we will see how the ideas of frontier orbitals can be applied to cycloaddition reactions.

7.1.1 Cycloadditions

A simple (though not very efficient or convenient) cycloaddition reaction is the reaction of buta-1,3-diene with ethene to give cyclohexene (reaction 7.3). The reaction does not appear to involve significant charge build-up in the transition state, and it appears that both new bonds are being formed at the same time, since the reaction is stereospecific with *cis* addition across both unsaturated systems.

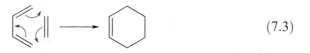 (7.3)

It can be seen in Figure 7.6 that there is favourable (same phase) overlap between both ends of the HOMO of the butadiene molecule with the LUMO of the ethene molecule, and *vice versa*. Accordingly, as the molecules approach, there is a favourable energy change and the reaction takes place readily.

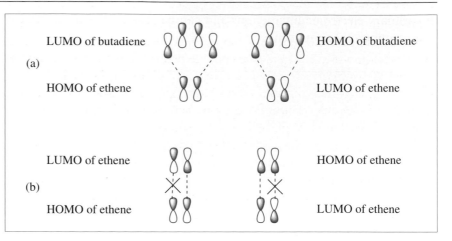

Figure 7.6 HOMO–LUMO interactions for (a) ethene with buta-1,3-diene and (b) ethene with ethene

However, for reaction (7.2), the dimerization of ethene to cyclobutane, the HOMO–LUMO interactions are unfavourable; if there is in-phase (favourable) interaction at one end, it will be unfavourable at the other. Thus the interactions cancel out, there is no overall favourable frontier orbital interaction, and the reaction does not take place thermally.

Other Cycloadditions

Provided that there is a favourable interaction at both ends between the HOMO of one component and the LUMO of the other component, reactions with a cyclic transition state should be possible for many other systems. Figure 7.7 shows the basic structures available for π systems containing between two and four overlapping p orbitals.

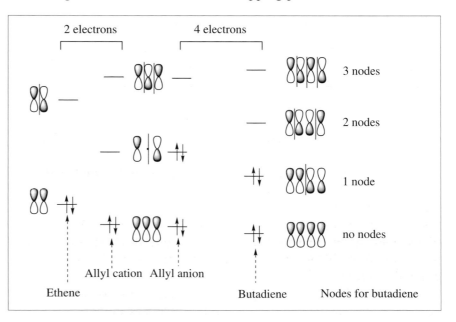

Figure 7.7 Molecular orbitals derived from two, three and four p orbitals. Nodes are shown in red. Orbital phases are shown as shaded and white

Inspection of Figure 7.7 shows that whatever the size of the system, the lowest, most bonding orbital has all the constituent orbitals in phase. Thus the components at the two ends necessarily have the same phase. For the next orbital up, there is a reversal of phase in the middle of the molecule, giving rise to antibonding at this point for systems with even numbers of orbitals, or non-bonding if there is an odd number of constituents. This change of phase or node results in the two end component orbitals being out of phase. For ethene (two atomic orbitals in $\Rightarrow$ two molecular orbitals out) there are no further orbitals, but for systems with three or more constituent orbitals the next orbital up in energy will have two nodes, and the end orbital components will be back in phase again. This alternation of sign will continue as we go to higher and higher energies. Since electrons go into the lowest orbital available, a system with two electrons will have its HOMO with the end component orbitals in phase irrespective of the number of atomic orbitals involved. For four electrons, the HOMO will have the end components out of phase, for six, they will be back in phase, and so on. Thus for systems with $4n+2$ electrons the end components will be in phase, and for $4n$ electrons the ends will be out of phase. Since the LUMO is the next orbital above the HOMO, the phases will be opposite, with $4n+2$ electrons resulting in out-of-phase end components, and $4n$ in-phase. The position is summarized in Table 7.1, which enables the phases at the ends of molecular orbitals to be easily determined without having to construct all the molecular orbitals.

Table 7.1 Phases at ends of frontier orbitals

Frontier orbital	$4n+2$ electrons	$4n$ electrons
HOMO	Same phase	Opposite phase
LUMO	Opposite phase	Same phase

The atoms need not be carbon: it is the orbitals that matter. Charged species can be involved, for example the allyl cation $CH_2=CH-CH_2^+$ has two π electrons (formally two from the double bond and none from the carbocation centre), the allyl anion $CH_2=CH-CH_2^-$ has four π electrons (two from the double bond and two from the lone pair on the negatively charged carbon). Ozone, $O=O^+-O^-$, is isoelectronic with the allyl anion, and will also bring four π electrons to the reaction (two from the double bond and two from the lone pair on the negatively charged oxygen).

For conjugation to be effective in π systems, approximate planarity needs to be maintained so that each p-type orbital can overlap with the next. The two points of attack on the π system can be on the same surface (top or bottom), in which case the attack is said to be **suprafacial** (supra = above) for that π system (Figure 7.8). Less commonly, attack can take place on opposite surfaces, in which case it is said to be **antarafacial** (possibly from antae, antarum = columns on either side of a door).

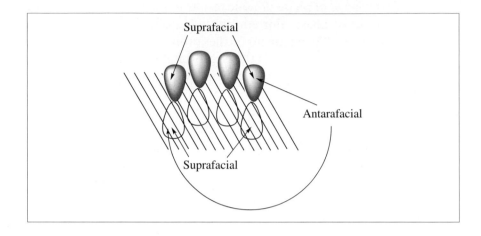

Figure 7.8 Suprafacial and antarafacial attack

Suprafacial cycloadditions are favourable if 4n+2 electrons are involved.

Almost all cycloadditions are suprafacial for both components. For these fully suprafacial reactions, inspection of Table 7.1 shows that for the ends of the HOMO and LUMO to both be in phase (or both to be out of phase) and therefore to give favourable interactions at both ends, one component needs $4x$ electrons and the other needs $4y + 2$, *i.e.* a total of **$4n + 2$ electrons** ($n = x + y$).

In a very small number of reactions, the twisted geometry of one of the components makes it possible for antarafacial attack to take place on one of the components. In these rare cases, since the opposite lobe of the antarafacial component is being used, the $4n + 2$ rule is broken, and $4n$ electrons becomes the favoured total. Heptafulvalene (**2**) has a twisted structure (Figure 7.9) and reacts antarafacially with tetracyanoethene to give the adduct **3** (reaction 7.4).

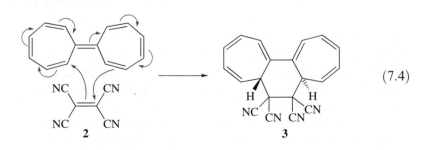

(7.4)

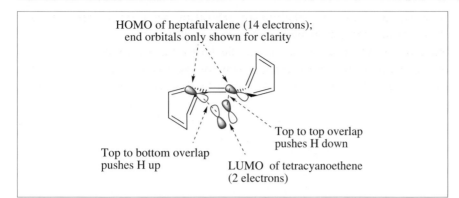

HOMO of heptafulvalene (14 electrons);
end orbitals only shown for clarity

Top to top overlap
pushes H down

Top to bottom overlap
pushes H up

LUMO of tetracyanoethene
(2 electrons)

Figure 7.9   Antarafacial attack of tetracyanoethene on heptafulvalene

7.2   The Aromatic Transition State

The frontier orbital arguments outlined above strictly apply only to the initial stages of the reaction when the molecules come into contact and the orbitals start interacting. We have shown that there will be favourable HOMO–LUMO interactions if $4n + 2$ electrons are involved in a fully suprafacial reaction or $4n$ electrons if one antarafacial component is involved. It does not necessarily follow from this that there will be a lowering of energy in the transition state. *However, in the transition state of all fully suprafacial pericyclic reactions, there is a cyclic overlap of the <u>atomic</u> orbitals involved in the bonds being formed and broken in the reaction, and we may expect, by analogy with cyclic conjugated structures, that delocalization and hence stabilization will be involved if there are the correct number of electrons.*

7.2.1   Hückel Systems

For all common aromatic systems, the overlap of the constituent p-type atomic orbitals is continuous around the cyclic system (no phase change). This leads in every case to a uniquely stable orbital with in-phase overlap between each of the adjacent orbitals in the cyclic system, as shown for three-, four-, five- and six-membered ring systems in Figure 7.10. This orbital can accommodate two electrons. The next most-stable orbitals will have a node where the phase changes, causing antibonding across this plane. Because the molecule is cyclic, there are *two* ways of introducing this nodal plane, at right angles to each other. This results in two orbitals of equal energy, which can between them accommodate four electrons. For the larger ring sizes, the higher orbitals will have two, three or more nodes, still occurring in pairs which can accommodate four electrons. Thus aromatic systems will have $4n + 2$ electrons, the **Hückel number**. Figure 7.10(b) and (d) shows that, for even numbered ring sizes, there is a unique most antibonding orbital with out-of-phase overlap between all

For suprafacial pericyclic reactions, the cyclic transition state is stabilized if a Hückel number of electrons, $4n + 2$, is involved. This will give faster reaction rates.

next-neighbouring orbitals, but an orbital of this type is impossible for rings with odd numbers of atoms as in (a) and (c). For a three-membered ring, the aromatic number of two electrons can be achieved by loss of an electron from the neutral cyclopropenyl radical $C_3H_3^\bullet$ to give the cyclopropenyl cation $C_3H_3^+$ (**4**). For a five-membered ring, six electrons can be achieved by the neutral cyclopentadienyl radical $C_5H_5^\bullet$ gaining an electron to give the cyclopentadienyl anion $C_5H_5^-$ (**6**), or in the case of pyrrole (**7**) by the contribution of two electrons from the lone pair on the nitrogen atom. Cyclobutadiene (**5**) should be a triplet with an electron in each of the equivalent orbitals if the molecule is delocalized. In fact, molecules of this type distort to give alternate single and double bonds. Benzene (**8**) is the archetypal aromatic molecule, with an unstrained ring system with sp^2 hybridization, 120° bond angles and six π electrons that can be accommodated in the three bonding molecular orbitals.

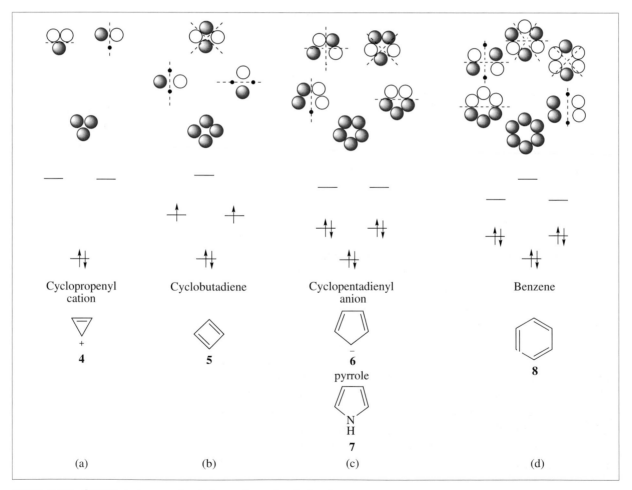

Figure 7.10 Orbitals and energy levels for cyclic conjugated systems

7.2.2 Möbius Systems: Rings with a Twist

In a large cyclic conjugated system, we could envisage a small angular deviation between the p orbitals of adjacent atoms, which if continued in the same direction round the ring could eventually reach a 180° total as we get back to the original atom, giving an out-of-phase antibonding interaction at this point. Systems of this type are rare, but it has been suggested that the cyclononatetraenyl cation $C_9H_9^+$ (**9**) may be a system of this type (Figure 7.11). They are named **Möbius systems** after the topological structure known as the Möbius strip, which can be constructed by joining the ends of a long thin strip of paper into a loop, but turning one end through 180° before joining.

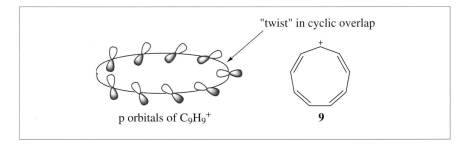

"twist" in cyclic overlap

p orbitals of $C_9H_9^+$ **9**

Figure 7.11 Orbital overlap in a Möbius system

Since there is a "twist" in the system, there is no uniquely favourable orbital with positive overlap all the way round the ring, a crucial difference from Hückel molecules. The change of phase at a position on the opposite side of the molecule to that of the first molecular orbital (Figure 7.12a) will result in another orbital (Figure 7.12b) with positive overlap between all but one pairs of adjacent atoms, and with the same energy. These two orbitals will accommodate four electrons. We can put further nodes in the system, always in pairs, giving a succession of pairs of orbitals which can accommodate four electrons. Finally, if there is an odd number of atoms in the ring, there will be an unique high-energy orbital, with out-of-phase overlap between all pairs of adjacent orbitals. Since the orbital levels occur in pairs which can accommodate four electrons, Möbius systems will be stabilized if they have **4n electrons** and destabilized with $4n+2$, the opposite to Hückel systems. The molecular orbitals for a hypothetical Möbius cyclobutadiene are shown in Figure 7.12; the generalized progression of orbital levels is shown for both Hückel and Möbius systems in Figure 7.13.

Although Möbius molecules are rare, Möbius transition states, with one antarafacial component, are relatively common. For these systems, 4n electrons is the favourable number, giving a stabilized transition state and a faster reaction rate.

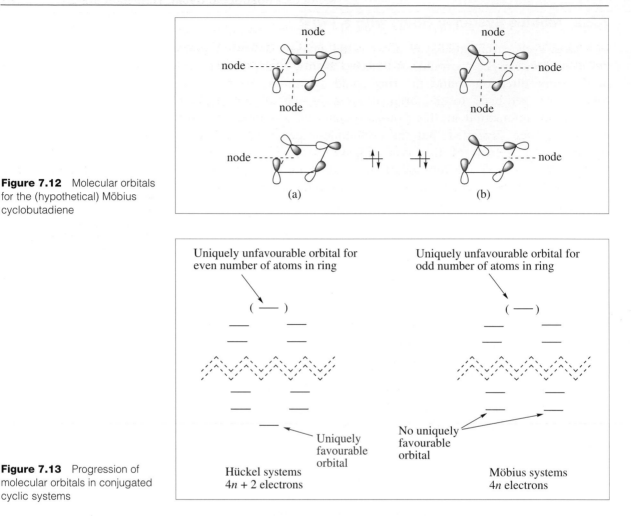

Figure 7.12 Molecular orbitals for the (hypothetical) Möbius cyclobutadiene

Figure 7.13 Progression of molecular orbitals in conjugated cyclic systems

7.2.3 The Link between the "Frontier Orbital" and the "Aromatic Transition State" Approaches

We have established earlier in the chapter that there will be favourable Frontier Orbital HOMO–LUMO interactions when two molecules approach for a cycloaddition reaction if there are $4n + 2$ electrons involved in a fully suprafacial reaction, or $4n$ electrons if there is an antarafacial component. For delocalization of electrons in the transition state, the fully suprafacial cycloaddition reaction will result in a continuous cyclic overlap of atomic orbitals in the transition state without a phase change, for which $4n + 2$ electrons will give aromatic stabilization. For a cycloaddition with one antarafacial component, the cyclic overlap of orbitals will give a Möbius system for which $4n$ electrons will provide stabilization. Thus the two approaches, Frontier Orbitals and the Aromatic Transition State will always be in agreement: favourable

HOMO–LUMO interactions as the molecules approach will always be accompanied by a lowering of the energy of the transition state, corresponding to a much faster reaction, as shown in Figure 7.14.

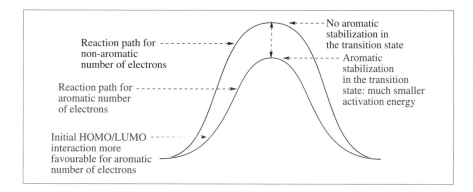

Figure 7.14 Reaction paths for pericyclic reactions involving aromatic and non-aromatic numbers of electrons. Number of aromatic electrons is $4n + 2$ for Hückel systems with no antarafacial components, or $4n$ electrons for Möbius systems with one antarafacial component

7.3   Application of the Idea of the Aromatic Transition State to Pericyclic Reactions

In the remainder of this chapter we will consider further cycloaddition reactions and other examples of pericyclic reactions. We will use the aromatic transition state approach for simplicity, although in all cases an approach based on HOMO–LUMO interactions would give the same result.

7.3.1   Cycloadditions

Entirely Suprafacial Reactions with $4n + 2$ Electrons

Diels–Alder reactions (for example 7.1 and 7.3) are important six-electron systems which can be used synthetically. The six-membered transition state is unstrained, and the probability of both molecules colliding with the correct orientation is reasonably favourable.

Reactions involving a five-membered transition state are even more favoured on entropy grounds, although there is some angle strain in the transition state and the products. The first step in the reaction of ozone with alkenes is a reaction of this type (7.5), as is the reaction of diazomethane with alkenes such as methyl acrylate (methyl propenoate) (reaction 7.6). In each case, six p-type electrons are involved, a Hückel number, so the reactions will be favoured as long as the reactions are entirely suprafacial. The movement of three pairs of electrons (six in total) is shown by the curved arrows, and corresponds to the bonds broken and formed during these two reactions.

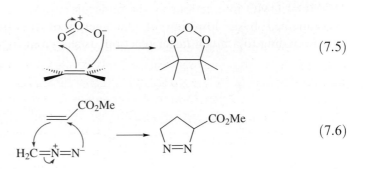

(7.5)

(7.6)

Reaction (7.7) involves the reversible removal of a proton from an allylic position by a base to give the stabilized anion **10,** which then reacts suprafacially with *trans*-1,2-diphenylethene to give the cyclopentane product in which the relative stereochemistry of the two phenyl groups has been preserved.

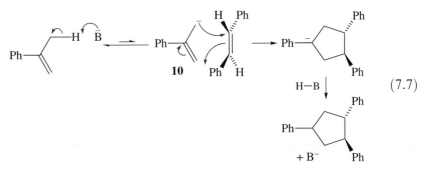

(7.7)

Suprafacial cycloadditions involving 10, 14, *etc.*, electrons are possible, but for open chain systems the probability of getting conformations of the two molecules in which interaction at both ends of both π systems takes place at once is very small, making the reactions impractical. However, if the molecules are constrained to give a favourable transition state geometry, for example by being part of a ring system, such reactions are feasible. In reaction (7.8), both components are constrained by being part of ring systems. The cyclic ketone **11** only brings six electrons to the reaction; the C=O group, though conjugated, does not react and is present in the product. The involvement of 10 electrons, a Hückel number, is shown by the five curved arrows.

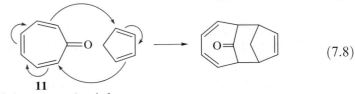

(7.8)

11
6 electrons not 8   4 electrons

Cycloadditions with an Antarafacial Component

These are rare. We have already met a 16-electron system in reaction (7.4), where the bicyclic geometry of the 14-electron system holds the ends of its π system close together and at an angle which facilitates antarafacial attack by the substituted ethene molecule.

Two ethene molecules do not react thermally to give cyclobutane (reaction 7.2). This 4π system has the wrong number of electrons for suprafacial attack to take place, and geometric reasons make it impossible for overlap to take place from a p-orbital lobe away from the direction of approach to give an antarafacial component (Figure 7.15).

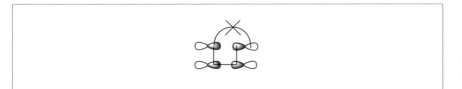

Figure 7.15 Impossibility of antarafacial attack in ethene dimerization

In principle, two buta-1,3-diene molecules could react to give cycloocta-1,5-diene (reaction 7.9), but this would involve an eight-membered transition state, with a low probability of the correct orientation for reaction. Instead, reaction can take place more readily via a six-membered Hückel transition state to give 4-vinylcyclohexene (reaction 7.1).

Mobius

One antarafacial
component

(7.9)

7.3.2 Electrocyclic Reactions

Pericyclic reactions can involve σ as well as π bonds. Electrocyclic ring opening (or closure) reactions are exemplified by reactions (7.10) and (7.11).

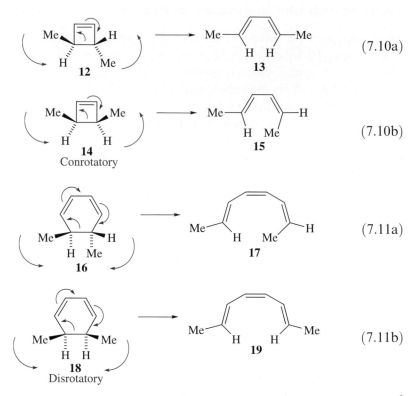

$$(7.10a)$$

$$(7.10b)$$

$$(7.11a)$$

$$(7.11b)$$

These reactions involve breakage of a σ bond. As the bond breaks, the sp^3 orbitals gradually rehybridize to p orbitals, and rotation takes place so that the incipient p orbitals can overlap with the p orbitals of the π system in the ring, resulting in one more double bond in the open-chain product. The reactions are stereospecific: *trans*-3,4-dimethylcyclobutene (**12**) gives *trans,trans*-hexa-2,4-diene (**13**), whereas the *cis* isomer (**14**) gives *cis,trans*-hexa-2,4-diene (**15**).

In reactions (7.10a) and (7.10b), the rotations are both in the same direction and are shown as anti-clockwise. For (7.10b), the rotations could have equally well been both clockwise, but for (7.10a), clockwise rotation would have given rise to the less stable *cis,cis* isomer which would be energetically less favourable. Where the rotations are in the same direction, the reaction is said to be **conrotatory**.

By contrast, the cyclohexadiene isomers in reaction (7.11) undergo ring opening with rotations in opposite directions as shown. These reactions are termed **disrotatory**.

These electrocyclic reactions and their conrotatory or disrotatory nature can be readily understood on the basis of aromaticity in the transition state (Figure 7.16). For the conrotatory mode, the rotations of the breaking σ orbitals bring about a phase change in the cyclic transition state: continuous red-to-red overlap cannot be maintained. Thus the transition state is Möbius, and is favoured for 4 (or any multiple of 4) electrons in the reacting system, as in Figure 7.16a. The distrotatory mode (Figure 7.16b) would have continuous in-phase overlap, giving a Hückel system, for which 4n electrons is unfavourable.

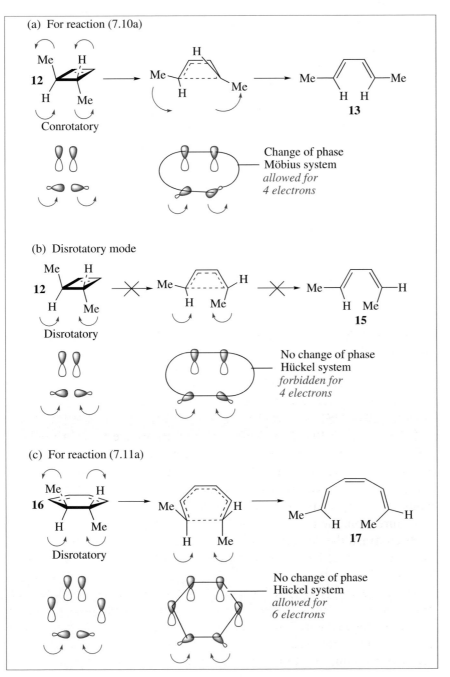

(a) For reaction (7.10a)

Conrotatory

Change of phase
Möbius system
*allowed for
4 electrons*

(b) Disrotatory mode

Disrotatory

No change of phase
Hückel system
*forbidden for
4 electrons*

(c) For reaction (7.11a)

Disrotatory

No change of phase
Hückel system
*allowed for
6 electrons*

Figure 7.16 Conrotatory and disrotatory ring opening of cyclobutenes and cyclohexadienes

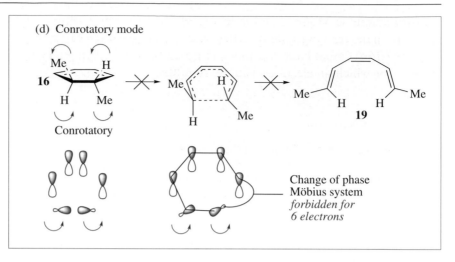

(d) Conrotatory mode

Figure 7.16   (Continued)

In contrast, ring opening of the cyclohexadiene isomers **16** and **18** to give **17** and **19** are both disrotatory. For this reaction, with a Hückel transition state involved, $4n + 2$ electrons is the favoured number, and for this number of electrons the conrotatory mode which would give a Möbius transition state is not favoured.

A worked problem follows and several examples of electrocyclic reactions occur in the problems at the end of the chapter. The important principle is that for $4n$ electrons, conrotation will give the favoured Möbius transition state, whereas for $4n + 2$ electrons, disrotation will give the favoured Hückel transition state.

Electrocyclic ring opening and closure will be disrotatory for $4n + 2$ electrons, but conrotatory for $4n$ electrons. Many examples of both types are known.

Worked Problem 7.2

Q   On heating, the aziridine **20** undergoes ring opening to the dipolar product **21** in reaction (7.12). Explain why the conrotatory mode shown is favoured.

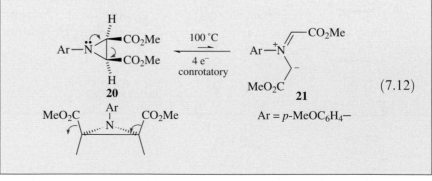

$$\tag{7.12}$$

A Four electrons are involved in this ring-opening reaction: two from the lone pair on the nitrogen and two from the σ bond which breaks, as shown by the curved arrows. Thus a Möbius transition state is favoured, resulting in conrotatory ring opening.

7.3.3 Sigmatropic Rearrangements

In these reactions, a σ bond moves across a π system to a new site. An example is reaction (7.13), where the stereochemistry of the product **22** shows that the reaction is suprafacial. The reaction is designated [1,5] because a σ bond to a single atom moves from one end to the other of a 5-atom system. Another interesting example of this is the rearrangement and interconversion of the methylcyclopentadiene isomers **23**, **24** and **25**, which is rapid at room temperature (reaction 7.14).

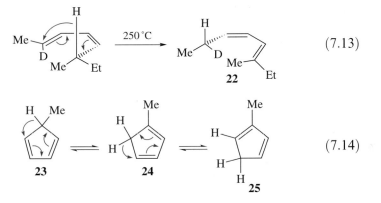

$$(7.13)$$

$$(7.14)$$

Figure 7.17 shows the cyclic overlap of the orbitals for a [1,5] suprafacial shift. This will be favourable for Hückel systems with $4n + 2$ electrons, as in reactions (7.13) and (7.14). Antarafacial migration will be possible for systems with $4n$ electrons; these are relatively uncommon. Hydrogen atoms do not migrate across an allyl system (reaction 7.15). The suprafacial migration is forbidden; the allowed antarafacial process cannot take place for geometric reasons which prevent the hydrogen orbital from overlapping simultaneously with the orbitals on carbon atoms 1 and 3 in an antarafacial manner.

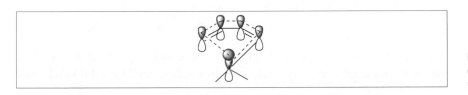

Figure 7.17 Cyclic overlap of orbitals in a [1,5] sigmatropic shift

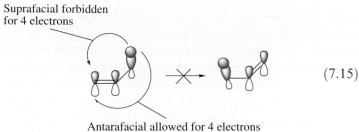

Suprafacial forbidden
for 4 electrons

Antarafacial allowed for 4 electrons
but sterically impossible in a short chain

(7.15)

Sigmatropic rearrangements will
be suprafacial for $4n + 2$
electrons, but antarafacial for
$4n$ electrons. Antarafacial
reactions are rarer, since
antarafacial movement is often
impossible geometrically.

A number of [3,3] rearrangements are interesting, including the rearrangement of allyl phenyl ethers **26** to dienones such as **27** (reaction 7.16). The designation [3,3] indicates that there are two 3-atom components of the system, shown in red in (7.16) and (7.17), which are joined at one end in the starting molecule, and at the other end in the products. The Cope rearrangement of **28** to **29** (reaction 7.17) is another example of a [3,3] rearrangement.

(7.16)

26   **27**

(7.17)

28   **29**

7.4   Photochemical Reactions

All the examples so far in this chapter have been thermal reactions. Many reactions which do not go thermally, such as (7.2), can be made to take place photochemically. If the alkenes are disubstituted, the stereochemistry of the products is consistent with entirely suprafacial attack, as shown in reaction (7.18).

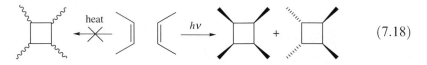

(7.18)

Frontier orbital treatment provides a good explanation. For the thermal reaction of two ethene molecules, HOMO–HOMO and

LUMO–LUMO are unable to contribute stabilizing influences as the reagents collide, as shown in Figure 7.18(a). For this four-electron system, HOMO–LUMO interactions are also unfavourable because the orbitals at the ends of the two π systems are out of phase. In the photochemical reaction, one of the ethene molecules has an electron promoted to a higher orbital, usually from the HOMO to the LUMO. This dramatically alters the possible interactions, as shown in Figure 7.18(b). We designate as "HOMO" and "LUMO" the orbitals that were the HOMO and LUMO before excitation. The orbitals themselves will be virtually unchanged by the change in occupancy. The "HOMO" with one electron will be able to interact with the HOMO of another molecule (two electrons) to give bonding and antibonding combinations of molecular orbitals. Two electrons will go into the bonding orbital, one in the antibonding, giving a net bonding interaction. Likewise the "LUMO" with one electron interacts with the LUMO of the second molecule (no electrons) to give bonding and antibonding combinations. The single electron available goes into the bonding combination with a reduction in energy. Thus for this system, which is thermally forbidden, both "HOMO"–HOMO and "LUMO"–LUMO reactions are favourable and the reaction is photochemically allowed.

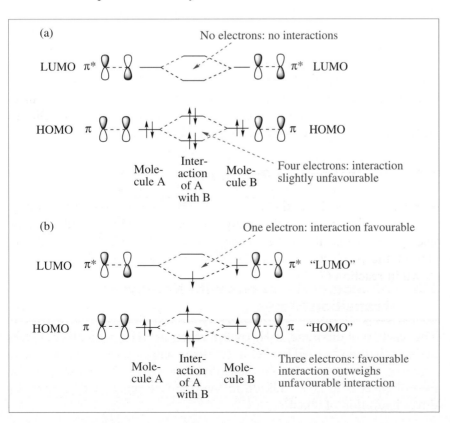

Figure 7.18 Molecular orbitals involved in thermal and photochemical reactions (a) Thermal: overall interaction slightly unfavourable. (b) Photochemical: one electron in molecule B promoted to "LUMO"; overall interactions favourable

The conclusion is general: reactions which are forbidden thermally are allowed photochemically and *vice versa*. For a particular number of electrons, cycloadditions that go entirely suprafacially under thermal conditions will require an antarafacial component photochemically and *vice versa*. For electrocyclic reactions that are conrotatory under thermal conditions, the corresponding photochemical reactions will be disrotatory and *vice versa*.

The **Woodward–Hoffmann rules** are a general expression of this. *A pericyclic reaction which is entirely suprafacial is allowed thermally if 4n + 2 electrons are involved (Hückel transition state), but forbidden for 4n electrons. If there is one antarafacial component, the reaction will be allowed thermally if 4n electrons are involved (Möbius transition state), but forbidden for 4n + 2 electrons. For photochemical reactions, these rules are reversed.* Roald Hoffmann shared the Nobel prize for Chemistry with Kenichi Fukui in 1981 for his contribution to this concept; Robert Burns Woodward had already won the prize in 1965.

The different stereochemistry of thermal and photochemical reactions can sometimes be exploited, where reactions are reversible, to cause isomerization between stereoisomers. The thermal conrotatory ring opening of *trans*-3,4-dimethylcyclobutene (**30**) to *trans,trans*-hexa-2,4-diene (**31**) goes effectively to completion, since the product **31** is more stable because of the relief of the ring strain (reaction 7.19). The photochemical ring closure is also very effective: the diene **31** because of its conjugation absorbs light at longer wavelengths than the mono-ene product *trans*-3,4-dimethylcyclobutene (**32**) which is photochemically inert to normal UV radiation, and the photon of light provides plenty of energy to make the endothermic reaction proceed. By this sequence of reactions, **30** can be transformed into its geometric isomer **32** in good yield.

For photochemical reactions, HOMO–"HOMO" and LUMO–"LUMO" interactions dominate, in contrast to the HOMO–LUMO interactions involved in thermal reactions. The rules for numbers of electrons involved in photochemical pericyclic reactions are the reverse of those for thermal reactions.

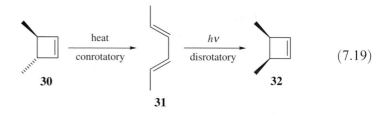

$$(7.19)$$

7.5   Molecular Reactions with Non-cyclic Transition States

The reaction of diborane, B_2H_6 (which reacts as BH_3), with alkenes gives adducts that can be used further in synthetic schemes, for example to give alcohols; this is important synthetically. Reaction (7.20) shows that the overall effect is to give addition in the opposite sense to that given by direct hydration of alkenes.

$$CH_3-CH=CH_2 \xrightarrow{BH_3} CH_3-CH_2-CH_2-BH_2 \longrightarrow (CH_3-CH_2-CH_2)_3B$$

$$\Big\downarrow \begin{array}{l} ^-OH \\ H_2O_2 \end{array} \quad (7.20)$$

$$CH_3-CH_2-CH_2-OH$$

The B–H bond has a polarity which is reversed compared with most H–X bonds, with the $\delta+$ on boron and $\delta-$ on the hydrogen atom. However, the reaction does not seem to be a normal electrophilic addition to the C=C double bond. Addition is suprafacial, suggesting a molecular reaction, which at first sight might appear to be a forbidden four-electron process. The movement of electrons shown in Figure 7.19(a) would indeed be forbidden, but there is an alternative involving the empty p orbital on boron. The movement of electrons shown in Figure 7.19(b) shows overlap between the sp^2 hybrid orbital on boron with the hydrogen, thence to the p orbitals on the two carbon atoms and back to the vacant p orbital on boron. However, there is no overlap between the p orbital on boron with the original sp^3 hybrid orbital; these are strictly orthogonal (no overlap), and hence there is no cyclic conjugation in the transition state. Thus the reaction is a non-cyclic concerted reaction involving the flow of two electron pairs.

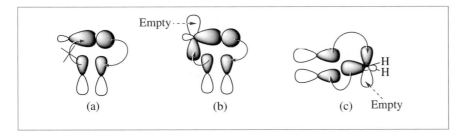

(a) (b) (c) Empty

Figure 7.19 Non-cyclic transition states

A similar situation occurs in the addition of carbenes to alkenes to give cyclopropanes, another synthetically useful reaction shown in its simplest form in reaction (7.21). Figure 7.19(c) shows the movement of electrons for this reaction, which is similarly concerted but does not involve a cyclic transition state.

$$\| + CH_2 \longrightarrow \triangleright \quad (7.21)$$

As a footnote, the stability and integrity of ordinary organic compounds containing nitrogen, oxygen and halogen atoms towards direct exchange of groups depends on the absence of empty accessible orbitals. Thus reactions of types (7.22) and (7.23), for example, which would have to involve a cyclic Hückel transition state with four electrons, do not take place. Peptides do not exchange side chain groups directly, and esterification of carboxylic acids has to take place by lengthier processes involving intermediates, as discussed in previous chapters.

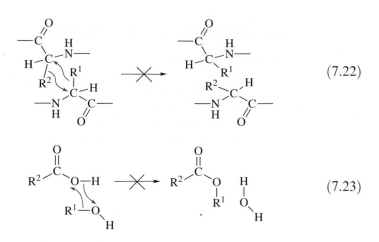

$$(7.22)$$

$$(7.23)$$

Summary of Key Points

1. The concepts of frontier orbital HOMO–LUMO interactions, the idea of an aromatic transition state, and the alternative concept of conservation of orbital symmetry (not developed in this chapter) all lead to the same result: for pericyclic reactions which involve a cyclic overlap of orbitals in the transition state, thermal reactions are allowed for reactions involving $4n+2$ electrons in Hückel systems (no change in phase between overlapped orbitals in the cyclic transition state) or for $4n$ electrons in Möbius systems (phase between overlapped orbitals in the cyclic transition state changes once on going round the ring). For photochemical systems, these rules are reversed.

2. The impossibility of suprafacial four-electron pericyclic reactions explains why simple interchange of groups between organic molecules does not readily take place, and why many reactions take place *via* intermediates, as described in earlier chapters.

Further Reading

I. Fleming, *Frontier Orbitals and Organic Chemical Reactions*, Wiley, London, 1976, Chapters 4 and 6.

R. B. Woodward and R. Hoffmann, *The Conservation of Orbital Symmetry*, Verlag Chemie, Weinheim, 1971.

M. Mauksch, V. Gogonea, H. Jiao and P. von R. Schleyer, $(CH)_9{}^+$, in *Angew. Chem. Int. Ed. Engl.*, 1988, **37**, 2395.

Problems

7.1. Using arguments based on frontier orbitals, show why two molecules of ethene do not react thermally to give cyclobutane, but ethene will react with buta-1,3-diene to give cyclohexene.

7.2. Using the concept of the aromatic transition state, show that, for pericyclic thermal reactions, $4n + 2$ electrons need to be involved if the reaction is entirely suprafacial, whereas $4n$ electrons are needed for reactions with one antarafacial component.

7.3. Using the formation of cyclobutane from ethene as an example, show that pericyclic reactions that are forbidden thermally are allowed photochemically and *vice versa*.

7.4. Classify the following six reactions as cycloaddition, sigmatropic, or electrocyclic. Use curved arrows to identify the number of electrons involved, and rationalize the results in terms of the Woodward–Hoffmann rules. If more than one possible product is listed, predict which of them will be formed. All reactions are thermal unless stated otherwise.

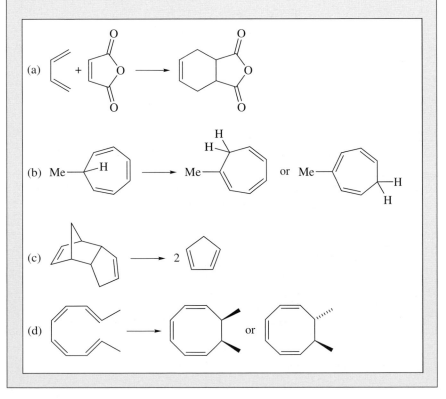

(e)   As (d) but photochemical

(f)

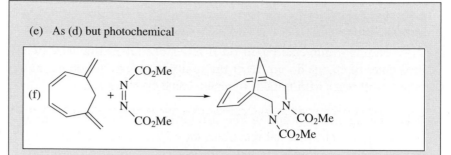

7.5.   What products would you expect from the photochemical reaction of *trans*-but-2-ene? [Hint: compare reaction (7.18)]

7.6.   On heating in CCl$_4$, the aziridine **A** slowly isomerizes to **B**, showing first-order kinetics, and giving eventually the equilibrium mixture. If dimethyl acetylenedicarboxylate (dimethyl butynedioate, **C**) is added, the isomerization does not take place, but instead the adduct **D** is formed. Explain these results. [Hint: consider reaction (7.12)]

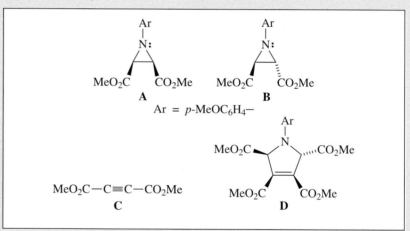

7.7.   Explain the formation of the cation **F** from the chloro compound **E** in the following reaction:

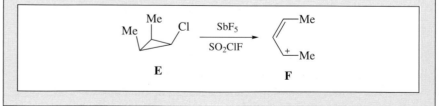

Answers to Problems

1.1.

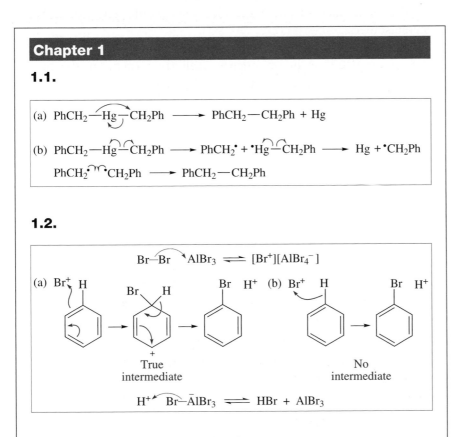

1.2.

1.3. (a) Unlikely. Reaction would involve four molecules interacting in one step. Also, the $Ni(CN)_2$ molecule is not used. A catalyst molecule must be involved in the reaction mechanism; if not, the reaction would go equally well in its absence.

(b) The reverse reaction would involve the interaction of five species: highly unlikely. Since the forward and back reactions of an elementary process must follow the same route, this reaction is unlikely because of the principle of microscopic reversibility. The reaction is more likely to proceed homolytically by steps in which one C-Pb bond is broken at a time.

(c) Quite plausible. Only two molecules are involved, and in the concerted process shown, three bonds are broken at the same time as three new bonds are formed, so the reaction will not be highly endothermic.

(d) Unlikely. If two bonds are formed simultaneously, they must be to the same face of the molecule. This would give the other stereoisomer.

(e) Unlikely. The nitrogen radical intermediate proposed would have nine electrons in its valence shell (two each for the four covalent bonds and one for the single electron that would make it a free radical).

1.4.

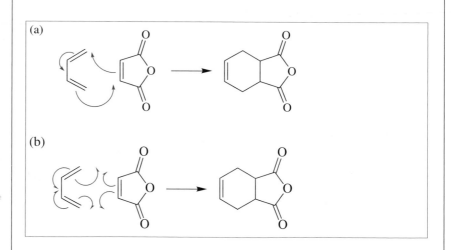

1.5.   The central carbon atom has four shared electrons in the covalent bonds to the two hydrogens, and two unshared electrons. The total number of electrons formally associated with the carbon in its valence shell is thus $(4/2) + 2 = 4$. This balances the $4+$ charge that a carbon atom with no electrons in its valence shell would have, making it electrically neutral.

Chapter 2

2.1. (a) Plots of [A], $\log_{10}$[A] and 1/[A] against time, made on a spreadsheet, are shown below. Because of the relatively small range of concentrations measured and the scatter of points, it is impossible to be sure which plot gives the best straight line.

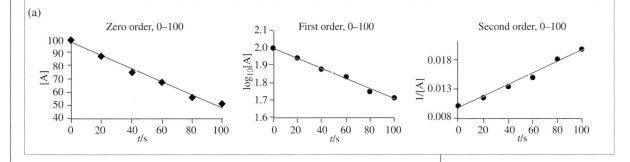

(b) The extended data are plotted below. The zero- and second-order plots both show pronounced curvature, whereas the first order plot is a much better straight line. These data support first-order dependence on [A], and show the importance of following a reaction for at least two half-lives if possible.

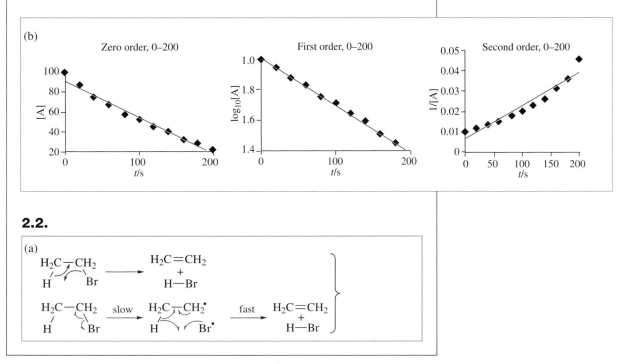

2.2.

(b)

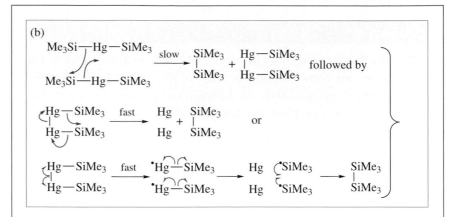

2.3. The nitric acid and sulfuric acid must react slowly to give an intermediate which reacts rapidly with benzene to form the product. Spectroscopic and other evidence shows this intermediate to be the nitronium ion, NO_2^+, formed by the reaction:

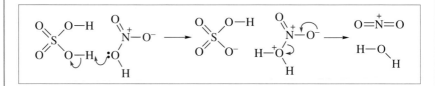

2.4. Since reaction p1 is fast, all $R^\bullet$ radicals, formed by reaction i, are rapidly converted into $ROO^\bullet$ radicals. Reaction p2 does not contribute to the loss of $ROO^\bullet$ radicals because they are immediately regenerated by p1. Thus in the steady state:

$$d[ROO^\bullet]/dt = 2k_i[\text{initiator}] - 2k_t[ROO^\bullet]^2 = 0$$

$$\therefore [ROO^\bullet] = (k_i/k_t)^{0.5}[\text{initiator}]^{0.5}$$

Product is only formed by p2, so:

$$d[ROOH]/dt = (k_i/k_t)^{0.5}k_{p2}[\text{initiator}]^{0.5}[RH]$$

Since the reaction of alkyl radicals with oxygen (p_1) is much faster than reaction p_2, most of the radicals present in the system at a particular time will be peroxyl radicals, $ROO^\bullet$, so the main route for destruction of the radicals will be the bimolecular reaction between the peroxyl radicals, *i.e.* k_t.

2.5.

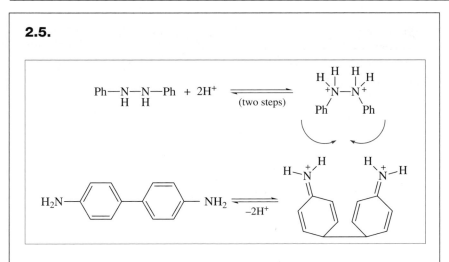

2.6. Mechanism (a) is a molecular reaction with a "tight" transition state, whereas the rate determining step in (b) is homolysis into two radicals with a "loose" transition state, which would give a larger Arrhenius A factor. The observed A factor falls within the range expected for molecular reactions shown in Table 2.1; thus (a) is favoured.

2.7. In the transition state of a molecular reaction, the molecules will have approached each other closer than the van der Waals contact distances, but the bond lengths of the bonds being formed between the terminal carbon atoms of the two molecules will still be much longer than typical C–C bond lengths, so there is significant bond breakage at these positions. However, the C2 and C3 atoms are changing their bonding, from single in one direction and double in the other, to the reverse arrangement in the product. The bond lengths will be intermediate between single and double, and little bonding will have been lost at these positions. Thus the observed isotope effect will be small, and in this example it is effectively unity.

2.8. Both mechanisms involve the breakage of C–H bonds in the transition state, so either would give significant hydrogen isotope effects, and no distinction between these mechanisms can be reached. However, in mechanism (a), no breakage of the C–N bond takes place in the transition state, whereas in the concerted reaction (b) the C–N bond is partially broken. Thus the nitrogen isotope effect favours route (b).

Chapter 3

3.1. The first reaction involves spreading a charge in the transition state; the second involves creation of charge. Thus the first will be decreased in rate by increasing the ionizing power of the solvent; the second will be increased (by a larger amount).

3.2. At the *meta* position, the inductive withdrawal of electrons by the oxygen atom outweighs the resonance release of electrons, so the net effect is electron withdrawal, shown by the small positive σ value. In the O–CO–Me group, there is a contribution to the structure from **L**:

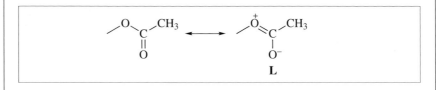

This produces a partial positive charge on the non-carbonyl oxygen of the group, which will make it more electron deficient. Hence it will withdraw electrons inductively more effectively than the OMe group, and will have a larger positive ρ value.

3.3. Examples:

σ_p^+	NH$_2$	NMe$_2$	OH	OMe	SMe	Cl
σ_p^-	CHO	COMe	CO$_2$Me	CN	NO$_2$	

3.4. Route (a) involves creation of positive charge in the intermediate formed in the rate-determining step, so ρ will be negative. The charge is conjugated with the aromatic ring, so σ^+ will be followed. Route (b) involves creation of negative charge in the intermediate formed in the rate-determining step, so ρ will be positive. The charge is not conjugated with the aromatic ring, so σ will be followed. Experimentally, ρ is found to be $+2.33$ and σ is followed, supporting mechanism (b).

3.5. The negative ρ value indicates that positive charge is being built up near the aromatic ring; the σ^+ dependence shows that this charge is conjugated with the aromatic ring and is therefore likely to be at the α-position, as shown below. The low ρ value indicates that the molecule is not fully ionized. A transition state of the type shown below is indicated.

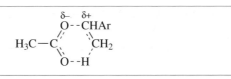

3.6. See Table 3.4, which defines the six categories. Choose other reactions whose mechanisms have been established and assign them to the six categories.

3.7. Introduction of the first *t*-butyl group only increases the rate by a factor of 1.2. The second *t*-butyl group increases the rate by the much larger factor of $18.4/1.2 = 15.3$. This increasing effect of successive substituents is characteristic of steric effects.

3.8. Introduction of the first phenyl group increases the rate by a factor of $5.2 \times 10^{-5}/1.76 \times 10^{-5} = 3.0$. The second phenyl group increases the rate by the smaller factor of $1.15 \times 10^{-4}/5.2 - 10^{-5} = 2.2$. This decreasing effect of successive substituents is characteristic of electronic effects.

3.9. Br and OH have the highest priority. Ethyl and methyl are both centred on carbon, but ethyl has the higher priority because of the second carbon atom attached to the first. The conformations shown here have the hydrogen atom of lowest priority at the back; if a structure were shown that did not have the hydrogen atom at the back, it would be necessary to rotate the molecule round any of the bonds except the C–H bond, to bring the H atom to the back. Going round the groups from Br > Et > Me shows an anti-clockwise sense for the isomer on the left; therefore this is the (*S*) enantiomer. In the product on the right, OH > Et > Me has a clockwise sense, so this is the (*R*) enantiomer.

3.10. All the L-α-amino acids shown in Figure 3.7 have NH_2 as the highest and H as the lowest priority groups. For almost all the α-amino acids, the CO_2H group takes precedence over the R group (also carbon atom centred) because of the two oxygen atoms attached to the carbon atom in the CO_2H group, giving the (*S*) configuration. In cysteine, $R = CH_2SH$, with a sulfur attached to carbon. Since sulfur has a higher atomic number than oxygen, CH_2SH will take priority over CO_2H. This reverses the priorities of these two groups, so the molecule will have the opposite config-uration, on this convention, *i.e.* (*R*).

3.11. In **X**, all the chlorine atoms are equatorial, so there is no hydrogen attached to an adjacent carbon atom in the *anti* conformation (the ring bonds are occupying this 180° dihedral angle orientation). Thus, with no favourable *anti* conformation available, reaction will be slow. The situation would not be improved if the molecule were to flip to the alternative chair conformation with all Cl atoms axial, since here all the *anti* positions would be occupied by chlorine atoms. **Y** has two axial chlorine atoms each with an adjacent axial hydrogen atom in an *anti* conformation, so reaction is faster. **Z** has two axial chlorine atoms, each with two adjacent axial hydrogen atoms in an *anti* conformation, so there are more possibilities for *anti* elimination, and reaction is faster still.

Chapter 4

4.1. See Sections 4.1, 4.2, 4.3 and 4.3.3 for definitions; alterna-tively, look in the index for relevant page numbers. The main types of reactions of carbanions and carbenes are listed in Section 4.3.3.

4.2. The three possible structures for the Ph_2CH^- anion with a negative charge on one of the benzene rings are shown as **b**, **c** and **d**. Similar structures can be drawn with the charge on the other ring, making seven structures altogether, including **a** which has the charge on the central atom. The Ph_3C^- anion has three benzene rings, hence 10 structures, three for each ring and one with the charge on the central carbon atom. Thus Ph_3C^- is more stabilized and Ph_3CH is more acidic than Ph_2CH_2. *Structures with alternative dispositions of the double bonds in uncharged rings will not contribute to extra stabilization of anions, since these structures are also available for the parent hydrocarbons.*

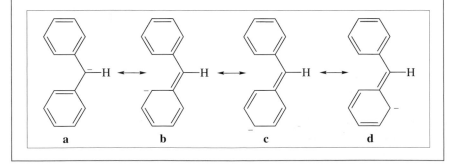

     a   **b**   **c**   **d**

4.3. Loss of a proton from CH_3CH_3 gives the alkyl anion $CH_3CH_2^-$. This has the charge on the not-very-electronegative carbon atom and no stabilization by delocalization. Thus CH_3CH_3 is an extremely weak acid. For CH_3COCH_3, the anion $CH_3COCH_2^-$ is delocalized, with one structure having the charge on the electronegative oxygen atom. Thus this anion is stabilized and CH_3COCH_3, though a weak acid, is much stronger than CH_3CH_3. CH_3CO_2Et has an intermediate acidity. Although the anion $^-CH_2CO_2Et$ has a structure analogous to that of $CH_3COCH_2^-$ with a negative charge on oxygen, the parent ester is stabilized by the structure $CH_3\overset{+}{C}(O^-)=OEt$ (similar to **9**), which decreases the *extra* stabilization in the ion.

4.4. (a) The first step involves abstraction of a proton from the chloronitrile by the strong base NH_2^-. The proton is abstracted from the carbon next to the nitrile group because the resultant anion is stabilized. The anion undergoes an internal nucleophilic substitution, closing the ring and expelling the chloride anion:

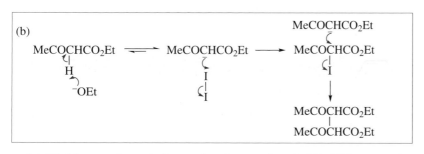

(a)

(b) The first step involves abstraction of a proton from the keto ester by the strong base EtO^-. As in (a) the proton is abstracted from the position which gives the most stable anion. The anion reacts by a S_N2 reaction with the iodine molecule to give the iodinated intermediate, which in turn undergoes a S_N2 reaction with the anion to give the dimeric product, with the displacement of an iodide anion:

(b)

$$MeCOCHCO_2Et \rightleftharpoons MeCOCHCO_2Et \longrightarrow MeCOCHCO_2Et$$

(c) Dichlorocarbene is formed as in reaction (4.35). This reacts by insertion into the reactive Si–H bond:

(c)

$$t\text{-}BuO^- + CHCl_3 \xrightarrow{(4.35)} Cl_2C: \longrightarrow H-CCl_2-SiEt_3$$
$$H-SiEt_3$$

4.5. (a) The three reactions prior to the rate-determining step are all reversible, so the hydroxynitrile anion **A** will be formed with an equilibrium concentration proportional to the two reagents, *i.e.* [PhCHO][CN$^-$]. The overall kinetics will be determined by the rate-determining step, with a kinetic dependence on [**A**][PhCHO] or [PhCHO]^2[CN$^-$], third order overall.

(b) The carbanion **A** is stabilized, with a structure involving negative charge on the nitrogen atom [compare Problem 4.4(a)]. No such stabilization is available for the corresponding hydroxide adduct, so although the hydroxide ion adds reversibly to the carbonyl group, an anion analogous to **A** cannot be formed, and the condensation cannot take place.

4.6. The reaction is likely to be a reverse of the aldol condensation (*e.g.* reaction 4.20). The first step involves equilibration of the keto alcohol with its anion. This is followed by a rate-determining fragmentation of the anion to acetone and the acetone anion, which will rapidly be converted by water into a second acetone molecule. Since the second step will be rate determining, the rate will be proportional to the concentration of the anion formed in the first step, which will be proportional to [OH$^-$][Me$_2$C(OH)CH$_2$COMe]:

$$Me-CO-CH_2-\underset{\underset{\displaystyle OH}{|}}{C}Me_2 \underset{H_2O}{\overset{^-OH}{\rightleftharpoons}} Me-CO-CH_2-\underset{\underset{\displaystyle \overset{|}{\overset{||}{O^-}}}{}}{C}Me_2$$

$$Me_2C=O \underset{^-OH}{\overset{H_2O}{\rightleftharpoons}} Me-CO-CH_2^- + O=CMe_2$$

Chapter 5

5.1. Electrophiles: HBr, I$_2$, H$_2$O, H$_3$O$^+$, NO$_2{}^+$, MeOH, AlCl$_3$, Me$_3$C$^+$, C$_6$H$_6$NO$_2{}^+$.
Nucleophiles: CH$_2$=CH$_2$, OH$^-$, H$_2$O, Me$_2$O, MeOH, PhO$^-$, C$_5$H$_5{}^-$.
Carbocations: Me$_3$C$^+$, C$_6$H$_6$NO$_2{}^+$.
Cations: Na$^+$, H$_3$O$^+$, NO$_2{}^+$, Me$_3$C$^+$, C$_6$H$_6$NO$_2{}^+$.

5.2. $Me_2(Ph)CBr$ will react more rapidly in S_N1 reactions. The intermediate tertiary carbocation is further stabilized by delocalization of the positive charge onto the *ortho* and *para* positions on the benzene ring, which will lower the activation energy required for its formation.

5.3. Reaction (a) involves the S_N1 hydrolysis of a tertiary bromide. The planar tertiary carbocation formed as an intermediate is stabilized by the three alkyl substituents. The S_N1 hydrolysis of 1-bromoadamantane in (b) would lead to a cation that could not have coplanar geometry at the carbocation centre. The cation would therefore have a high energy, and the reaction is very slow.

5.4. The presence in the products of compounds with both rearranged and unrearranged carbon skeletons suggests that a carbocation intermediate is formed. Anionic or radical intermediates do not normally rearrange. The mechanism is likely to involve the reaction of $PhCHMe–CHMeNH_2$ with HNO_2 to give $PhCHMe–CHMe^+$. This will give $PhCHMe–CHMeOAc$ by reaction with an acetic acid molecule. Rearrangement of the carbocation by a 1,2-hydrogen or -methyl shift will give the rearranged cations $PhMeC^+–CH_2Me$ or $Ph\overset{+}{C}H–CHMe_2$, respectively, which react with acetic acid to give $PhCMe(OAc)–CH_2Me$ and $PhCH(OAc)–CHMe_2$.

Carbocations rearrange readily, but carbanions and radicals do not rearrange so readily, so an intermediate of these types is much less likely. See Box 5.1. Anionic intermediates are very unlikely in acid solutions.

5.5. 1-Fluoropropane reacts with SbF_5 as shown below:

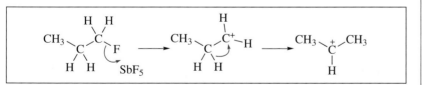

SbF_5 is a Lewis acid and can form a strong bond to another fluorine atom: the fluorine atom leaves the 1-fluoropropane with the electron pair from the bond, leaving behind the *n*-propyl (prop-1-yl) cation. This is a primary cation, which quickly rearranges by a proton shift to give the more stable prop-2-yl cation (a typical carbocation rearrangement). The secondary carbon bearing the formal charge is highly deshielding, so the attached proton shows its NMR peak at the very high δ value of 13.5 ppm. It is split into a septet by the six hydrogen atoms of the two adjacent methyl groups. The methyl hydrogens are not directly attached to a formally positively charged carbon atom, so the deshielding is less, but still significant. The peak is split into two by the single proton on the adjacent carbon atom. The intensity for the six protons is six times that of the single C–H proton. The coupling constants are the same because in a H_a–C–C–H_b system the influence of H_a on H_b through the two carbons is the same as that of H_b on H_a.

5.6.   Mechanisms:

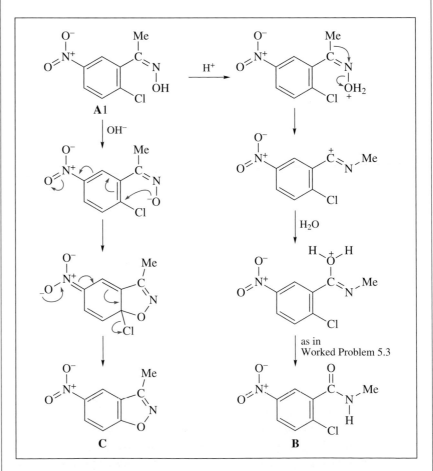

The oxime must have the structure **A1**, since ring closure takes place in the presence of base to give **C** by a nucleophilic substitution reaction. In the other isomer, the oxime oxygen atom would be too far away from the C–Cl bond for ring closure to be possible. This allowed the stereochemistry of these oxime isomers to be established long before X-ray analysis was available. The acid-catalysed rearrangement starts with the protonation of the oxime OH group. The resulting cation loses a water molecule in the second step. If the N–O bond were the only one to break, a very unstable NR_2^+ intermediate would be formed: most unlikely, and if it did, a mixture of amides would be expected from a symmetrical intermediate. Thus the methyl group migrates synchronously with the loss of the water molecule, giving the rearranged carbocation, which is converted by addition of water and proton shifts into the observed amide **B**.

This experiment shows that the group *trans* to the OH group in the oxime is the one that migrates in the Beckmann rearrangement.

5.7.   Mechanism:

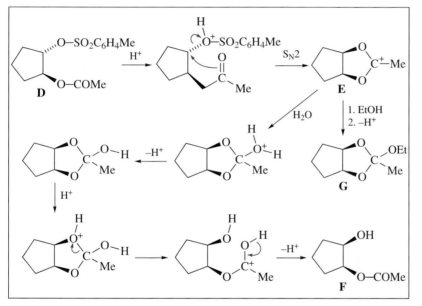

The first step is protonation. The second step is the displacement by the carbonyl oxygen of the $HOSO_2C_6H_4Me$ group by a S_N2 process to give **E**. Since it is an S_N2 reaction, the stereochemistry of the carbon is inverted. The new bond is formed from the top as the sulfonic acid leaves from the bottom. The remaining steps to give the product are typical of ester hydrolysis and involve no further possibility of stereochemical change on the cyclopentane ring, since no reaction takes place at either of the chiral carbon atoms.

To establish if the reaction is acid catalysed, carry out the reaction in the presence of different concentrations of acetic acid. If there is no acid catalysis, there will be no change in rate. If the rate is affected by acid concentration, try the reaction in buffer solutions of acetic acid/sodium acetate with a fixed ratio of [HOAc]/AcO$^-$]. This will keep [H$_3$O$^+$] constant, so if the reaction is specifically catalysed by H$_3$O$^+$, there will be no change in rate. If the increase in buffer concentration does cause an increase in rate, the reaction is subject to general acid catalysis, with a kinetic term involving [HOAc] as well as one involving [H$_3$O$^+$]. If ethanol is added to the mixture, **G** will be formed by the nucleophilic ethanol trapping the carbocation **E**.

Chapter 6

6.1.   Mechanism:

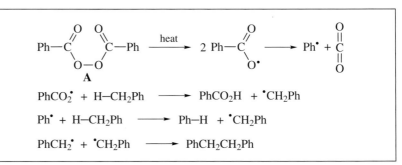

Homolysis of the weak O–O bond in the peroxide **A** gives two benzoyloxyl radicals. These will either fragment to give the phenyl radical and carbon dioxide, or abstract a hydrogen atom from the methyl group of a toluene molecule to give the stabilized benzyl radical and benzoic acid. The unstable phenyl radical will also readily abstract a hydrogen atom from toluene molecule to give the stabilized benzyl radical and benzene. The stabilized benzyl radicals have no radical-molecule reactions open to them, so they combine to give 1,2-diphenylethane.

6.2.   The following structures can contribute:

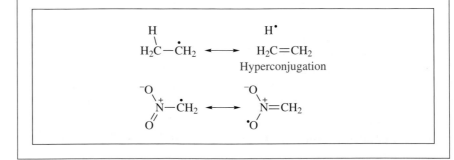

The diagrams refer to a substituted methyl radical, but similar stabilization would be available for any radical where the radical is conjugated with the reaction centre.

6.3. Although structures **B1** and **B2** can be written for the phenyl radical, two similar structures can be written for the benzene molecule, so there is no extra stabilization obtained when benzene loses a hydrogen atom to give the phenyl radical. No structures can be written where the unpaired electron is on a different atom, so the electron is localized in the sp^2 orbital on one carbon atom.

6.4. Mechanism:

$$PhCH_2{}^{\bullet} + {}^{\bullet}CH_2Ph \longrightarrow PhCH_2CH_2Ph$$

$$PhCH_2{}^{\bullet} + PhCH_2CH_2Ph \longrightarrow PhCH_2{-}H + Ph\overset{\bullet}{C}HCH_2Ph$$

$$PhCH_2{}^{\bullet} + Ph\overset{\bullet}{C}HCH_2Ph \longrightarrow \underset{\displaystyle PhCHCH_2Ph}{\overset{\displaystyle PhCH_2}{\overset{\displaystyle |}{}}}$$

Add 1,2-diphenylethane to the reaction mixture. This should increase the amount of 1,2,3-triphenylpropane produced. Alternatively, add a thiol or other radical trap. This should cut down the amount of both 1,2-diphenylethane and 1,2,3-triphenylpropane.

6.5. Mechanism:

$$t\text{-}Bu{-}O{-}O{-}Bu\text{-}t \longrightarrow 2t\text{-}Bu{-}O^{\bullet}$$

$$PhS{-}H + {}^{\bullet}O{-}Bu\text{-}t \longrightarrow PhS^{\bullet} + H{-}O{-}Bu\text{-}t$$

$$PhS^{\bullet} + CH_2{=}CMe_2 \longrightarrow PhS{-}CH_2{-}\overset{\bullet}{C}Me_2$$

$$PhS{-}CH_2{-}\overset{\bullet}{C}Me_2 + PhS{-}H \longrightarrow PhS{-}CH_2{-}CHMe_2 + PhS^{\bullet}$$

$\left.\right\}$ Propagation

6.6. The doublet is from the C–H at the radical centre, with a coupling constant similar to other alkyl radicals. The cyclohexyl radical will be expected to have the normal chair conformation, though with an sp^2 hybridized radical centre. A Newman projection (below) looking from C1 to either C2 or C6 will show an axial hydrogen nearly eclipsing the p orbital with the unpaired electron, giving a large coupling constant, and an equatorial C–H with a dihedral angle close to 90°, giving a very small coupling constant. There will be two equivalent hydrogens of both types, giving triplets.

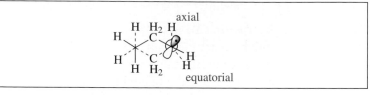

6.7. Abstraction of H4 in the cyclohexyloxyl radical from **C** would require the boat conformation **C′**, which is a very high-energy conformation. Cyclooctane is much more flexible, with considerable rotation round the C–C bonds, and conformations such as **E′** will be significant, allowing hydrogen abstraction to take place.

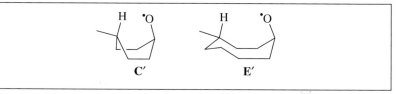

6.8. Mechanism:

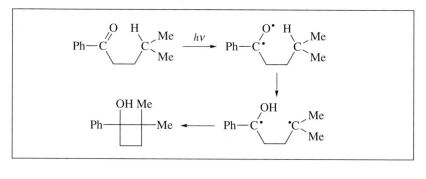

Chapter 7

7.1. Define frontier orbitals as in Section 7.1 and (referring to Figure 7.7 if necessary) establish that the HOMO of buta-1,3-diene has opposite phases at the end of its π system, as does the LUMO of ethene. Thus when these two molecules react in a cycloaddition, the phases match at both ends, giving favourable interactions. For two ethene molecules the HOMO has the same phase at the two ends but the LUMO has opposite phases, so there will be a favourable interaction at one end but unfavourable at the other.

7.2. In an entirely suprafacial system, there is positive overlap between all next-neighbour orbitals in the cyclic transition state. This will be delocalized, with a uniquely favourable molecular orbital with positive overlap all the way around the ring, which can hold two electrons. Higher orbitals occur in pairs (see Figures 7.10 and 7.13), holding four electrons in each level. Thus stabilization of the transition state (and rapid reaction) will take place if $4n+2$ electrons (the Hückel number) are involved. If there is a change in phase in the cyclic overlap (an antarafacial component), there is no uniquely favourable orbital, and since the orbitals occur in pairs with the same energy, $4n$ electrons will be favourable for these Möbius systems. See Section 7.2.

7.3. In thermal reactions, HOMO–LUMO interactions are the most important, for the reasons expounded in Answer 7.1. In photochemical reactions, since an electron has been promoted, HOMO–"HOMO" and LUMO–"LUMO" interactions will be favourable, and since the energy gap is small or zero, these interactions will dominate. Thus for reactions that have the wrong number of electrons for a thermal reaction, the photochemical reaction will take place; conversely, if the correct number of electrons for a thermal reaction is present, the photochemical reaction will not take place.

7.4. (a) Cycloaddition, six electrons, entirely suprafacial, thermally allowed:

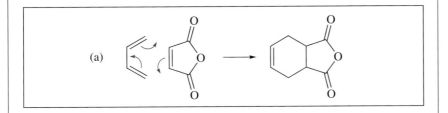

(b) [1,5] sigmatropic reaction. Six electrons, so suprafacial process involving migration of the hydrogen atom is thermally allowed:

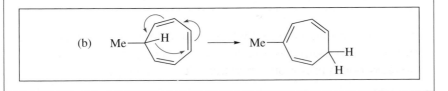

The other isomer would come from a [1,7] shift. This would be allowed for an antarafacial shift, but would be impossible geometrically.

(c) Reverse of cycloaddition, six electrons, suprafacial, so thermally allowed:

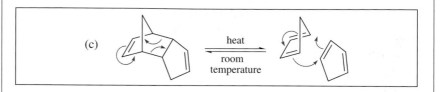

Proceeds to the right on heating (more entropy), but the reaction reverses at room temperature [favoured by energy (enthalpy)]. The pathway of an elementary reaction must be the same in both directions (microscopic reversibility).

(d) Electrocyclic ring closure. Eight electrons, so antarafacial component needed; hence conrotatory. Looking at the molecule from the right-hand side, both the terminal methyl groups point outwards. Conrotation in the clockwise sense will bring one up, the other down, as shown:

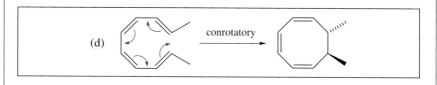

(e) Since the thermal reaction is conrotatory, the photochemical reaction must be disrotatory, giving the other (*cis*) isomer.

(f) Cycloaddition, 10 electrons, so suprafacial reaction allowed thermally:

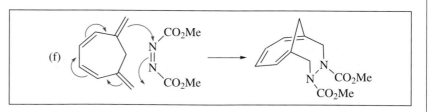

7.5.   This photochemical four-electron process is suprafacial. The two but-2-ene molecules can approach each other in two ways; the two new bonds are shown in red.

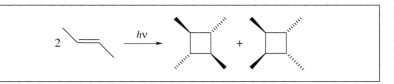

7.6. The aziridine **A** undergoes a reversible electrocyclic ring opening on heating. Since four electrons are involved, this is conrotatory. The equilibrium favours the starting compound **A**. The open-chain product **G** slowly isomerizes to **H** by rotation about a C–N bond. Since **G** and **H** are stabilized by delocalization, whereas the transition state between them (in which one of the C–N bonds has been rotated through 90°) is not, there will be a considerable activation energy for this isomerization. **H** will rapidly undergo ring closure to give **B**. If **C** is present, a rapid six-electron cycloaddition takes place. The suprafacial attack preserves the orientation of the CO_2Me groups, giving **D**:

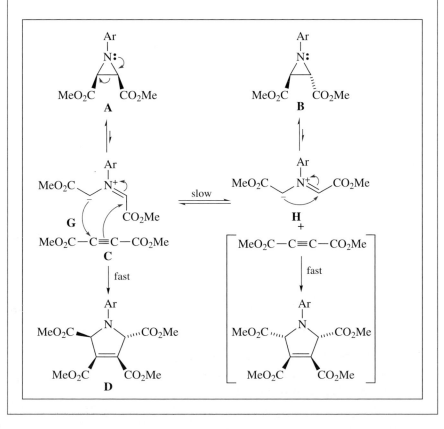

7.7.   In the polar aprotic solvent SO_2ClF, SbF_5 abstracts a chloride ion from **E** to give the cyclopropyl cation **G**. This rapidly undergoes electrocyclic ring opening to give the cation **I**, whose stereochemistry is established by NMR. Since two electrons are involved in the ring opening, the reaction is disrotatory:

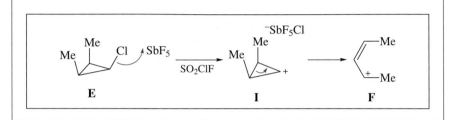

Subject Index

'Fashions fade, style is eternal.'

Yves Saint Laurent

1   Louise Henriksen design.

Contents

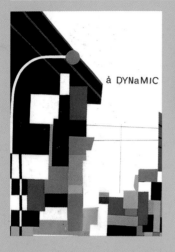

Textiles and Fashion

Contents

1 A design from Sandra
Backlund's Ink Blot Test
collection.

'*Working on fabrics, colour and inspiration, garments
go hand in hand at the beginning of a season as each
one inspires the other.*'

Michele Manz, senior director of womenswear for
Converse by John Varvatos

This book is for the textile designer who is interested in the
integration of textile design with fashion and also the fashion
designer who wants to fully integrate garment design with
textiles. Designers who will consider how the scale of a
design will work on the body, how the fabric will function
on the body through drape or structure, and how the fabric
will be cut and finished will benefit enormously from
reading this book.

The book endeavours to cover all the things you need to
know about fashion textiles. It begins with a brief history
of textiles, showing the links with technical innovation and
social developments. It then focuses on the processes of
textile design, including the ethical and sustainable issues
around textiles today. The book also provides practical
information on fibre production, dyeing and finishing
techniques. Also examined is how a fibre becomes a fabric
through construction techniques, for example, weave and
knit, and other more innovative processes. The book
continues by looking at the surface treatment of textiles
including print, embroidery and embellishment, and then
focuses on the way colour and trend can influence textiles
and fashion. The final section gives practical information
on the use of textiles within fashion design, how to choose,
cut and sew fabrics. Additionally, there is a very important
section on fashion and textile designers who work in the
industry, exploring what they do and how they use textiles
within their work.

All the text in this book is underpinned with visual examples
of fashion and textiles from designers who create wonderful
textiles. I hope their work will inspire you and that you gain a
1   great deal of pleasure from this book.

How to get the most out of this book

This book introduces different aspects of textiles and fashion via dedicated chapters for each topic. Each chapter provides numerous examples of work by leading fashion designers, annotated to explain the reasons behind choices made.

Key textiles and fashion principles are isolated so that the reader can see how they are applied in practice.

Clear navigation
Each chapter has a clear heading to allow readers to quickly locate areas of interest.

Designing textiles

1   Example of a mood board.

2–3   Examples of drawing techniques.

The next step is to collate the research that you have gathered. This gathering of informative textiles from classic textiles, cultures or other sources of inspiration can be in the form of mood boards or sketchbooks that document the research and create links to form a story that will develop into design ideas. It is important to then find a way to express your design ideas through drawing, collage, photography, or maybe CAD work. It is wise to also consider the surface you are going to design on: will you start to work on paper initially and then develop into cloth and knit or will you start to work directly with material? As you design you must understand the basic textile design principles of scale, texture, colour, pattern, repeat, placement and weight. Consider how these principles work within a sample and how these samples work together as ranges, as well as how your designs will result in functional, inspirational or commercial textiles suitable for use within contemporary fashion design and garment construction.

Rendering designs

You now need to think about what you are trying to design and how best to go about it. Determining the most appropriate medium to render your designs in is very important, whether it's paper, paint, pencils or a software package. Work out what is required and in what time frame. Bear in mind that you might need to learn new skills for the designs you are creating. Always remember to experiment and enjoy the process.

Drawing

Being able to communicate your ideas through drawing is fundamental to most design disciplines. However, it is possible to also use other media such as collage and photography as a means of communication. Experiment with drawing, use different types of media and be expressive with line, colour and texture. Think also about silhouettes and blocks of colour or tone within your design. Consider whether you are trying to represent what you are drawing precisely or if the artwork is developing in a more abstract direction.

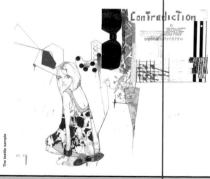

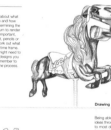

Researching textiles > **Designing textiles** > Textiles into production

---

Introductions
Special section introductions outline basic concepts that will be discussed.

Examples
Projects from contemporary designers bring the principles under discussion alive.

Textiles and Fashion

Additional information
Box-outs elaborate on subjects discussed in the main text.

Headings
These enable the reader to break down text and refer quickly to topics of interest.

Captions
These provide image details and commentary to guide the reader in the exploration of the visuals displayed.

Colour referencing

Colour often needs to be consistent across various fibres or fabric types, which in turn may require different types of dye that may even be produced in different countries. For a colour of a textile to remain consistent from the design stage through development to realisation, companies often use a colour referencing system. Pantone and the Munsell colour systems are common references for colour matching, as each colour has a specific number for reference. Rather than trying to describe the colour, the number can be used to identify the hue. Pantone charts are arranged chromatically by colour family and contain 1,925 colours. They are a great resource, but they are expensive and need to be replaced as the colours start to fade, making referencing inaccurate.

Looking at colour under different lighting conditions can affect the hue – an incandescent light places a yellow cast on the hue, while a halogen light creates a blue cast.

Colour and the customer

Colour is very important within fashion and textile design. When a customer enters a store they tend to be drawn to the colour of a garment. They may then go and touch the garment and lastly they will try it on to see if the fit is right.

Within a fashion collection safe colours are usually black, navy, white, stone and khaki. Buyers will often buy in garments in these colours as they are the staple colours of most people's wardrobes. It is sometimes a good idea to offer some of the basic colours and add to them seasonal experimental colours. These colours will add life to the collection and will ideally entice the customer to buy each season's new colours along with the trans-seasonal basics.

Skin tone can also have an effect on the colour choice of a garment. Dark skin looks great against strong, bright colours, while softer colours work better against paler skin.

1   A colour palette created by Justine Fox in response to the Chloé S/S08 collection. Copyright Global Color Research Ltd.

2   Chloé S/S08 runway show. Catwalking.com

3   Pantone colour book.

Khaki
During their years of colonial rule in India, The British Army dyed their white summer tunics to a dull brownish-yellow colour for camouflage in combat. This neutral tone was called 'khaki'. The word's origin is mid-19th century from the Urdu term kaki meaning 'dust-coloured' and from the Persian word kak, meaning 'dust'.

Season
Colours can also be seasonal. Cold seasons tend to warrant darker colours, such as blacks, browns and sludgy colours. As the season warms up the colours become lighter and paler. They then become stronger and brighter as the sun becomes more intense. The sun bleaches out pale colours, so if you are designing for hot countries consider a brighter colour palette. Think of the colour palettes of African textiles or Hawaiian shirts. When we pack for our summer holidays we quite often take brighter clothes than we would wear in a colder climate.

Chapter titles
These run along the bottom of every page to provide clear navigation and allow the reader to understand the context of the information on the page.

Running footers
Clear navigation allows the reader to know where they are, where they have come from and where they are going in the book.

How to get the most out of this book

'I get inspired by people, music, films, my own homes, travelling, the streets of London, Paris or New York. Great energy coming from meeting new and fun people, attending a great event, anything and everything feeds me in one way or another.'

Valentino in *Fashion: Great Designers Talking* by Anna Harvey

It is important to consider the function of the textile you are designing before you start. Is it required for its aesthetic qualities, how it drapes, the handle of the cloth, its texture, for its colour, pattern, surface interest, or is it required for its function, how it will stretch around the body or maybe how it can be tailored. Will it be used for its protective qualities, perhaps against rain or the cold? With the development of nano-textiles more advanced functions can be catered for – a fabric might deposit a medicine on the skin or be a form of communication, as the colour changes according to the wearer's temperature or mood.

It is useful to have knowledge of the historical development and use of textiles, for example, how different fabrics and techniques have become fashionable within Western fashion. It is also interesting to see how textiles are used in different cultures to clothe the body.

The inspiration for textile design can come from any source and it can inform colour, texture, pattern and scale. Consider the ways in which you might begin designing, what media you might use – paint, pencil, CAD – and what surface you might work on.

Once you have designed a range of textiles it is important to consider how you might sell your ideas or manufacture the design as a length of fabric or a garment.

Researching textiles

As with all designing it is important to look at what is happening in fashion and textiles currently (this is known as secondary research). This will enable you to direct your designs; do you want to do something similar to what is happening currently, to follow a trend and to be fashionable, or do you want to react against current ideas and try something more experimental and set a new trend or fashion?

Whatever you decide you will need to also find research that is original (known as primary research) in order for your designs to be new and not just copies of what is going on around you. Original research for textiles can come from anything: historical costume, galleries, nature, architecture, books, the Internet and travel, for example. It is important that your research can provide inspiration for imagery, pattern, texture, colour and silhouette.

A brief history of textiles

Toile peinte
This is hand-painted cloth.

Chint
A Hindu term for gaudily painted cloth that gave rise to the name 'chintz'.

Looking back historically we can see the types of textiles that were popular at certain times. This is usually related to some form of advancement in technology or trend within society.

Throughout the history of textiles, certain patterns and fabrics have been repeated. These textiles become classics and some classics remain constantly popular in some form or another, for example, spots, stripes and florals. Other classics go in and out of fashion, such as the paisley design. It is interesting to take a classic textile design and look at what makes it so timeless, then try to reinvent it.

1–2 Toile de Jouy designs originally depicted pastoral scenes that were finely rendered in one colour and positioned repeatedly on a pale background. In these examples, Timorous Beasties have taken the landscape of modern-day London to produce a contemporary toile de Jouy design.

1600s

1

2

The French government supported the development of the silk industry in Lyon. New loom technology and dyeing techniques were developed that produced fine-quality silks, surpassing the Italian silks, which had dominated the 16th century. The rococo period of the 17th century saw the fashion for very decorative dresses. An offshoot of this was chinoiserie, where designs were inspired by the cultures and techniques of the East. Patterns were asymmetric, many featuring oriental motifs, and were exotic in their colour combinations. Japanese kimonos became very popular and were imported by the Dutch East India Company. This company also imported from India a hand-printed cotton known as chintz. It was popular fabric as it was cheap, bright and colourfast. The popularity of the fabric threatened the French and British textile industries to such an extent that a ban on importing or wearing it was imposed.

1700s

1–2 A ladies' jacket from the 1800s. The fabric is tin-dyed black and lined with a small provençale cotton print.

In the early 1700s 'bizarre silks' were popular. The exotic plant shapes found on them were the result of the influence of Eastern culture. They made way for lace motifs, then large-scale luxurious florals in the 1730s, moving to smaller sprays of flowers.

In 1759 the ban on the cotton indiennes or chintz was lifted and the French textile industry again boomed. One factory in Jouy became famous for its printed cotton, the toile de Jouy.

Louis XV's mistress, Madame de Pompadour, wore a type of silk known as chiné à la branche or pompadour taffeta. The silk had a water-blotting pattern effect, which was achieved by printing the warp before weaving the fabric. During the 18th century England dominated men's fashion due primarily to its superior wool manufacturing industry and skilled tailors, while France dominated women's fashion.

At the end of the 18th century a simpler fashion to the rococo style became popular in women's clothing. A thin white cotton dress with little or no undergarments was worn, inspired by Greek and Roman antiquity. A muslin or gauze was best suited for this design as it offered a simple drape rather than moulding to the body. Cashmere shawls were worn over this garment in the winter. The shawls were brought back by Napoleon from his Egyptian campaign in 1799. The cashmere shawl came from the region of Kashmir in NW India. The wool of the mountain goat was spun into yarn to produce a light, soft, warm cloth of the highest quality. As a result these shawls were very expensive. By the 1840s the cashmere shawl had mass appeal and was made in small industries in France and Britain. Notably Paisley in Scotland produced a less expensive shawl and the pattern became associated with the region.

Jacquard
A fabric made on a jacquard loom. Named after French weaver and inventor, Joseph M Jacquard (1787–1834).

1800s

Once again the popularity of cotton in French fashion had grown to the point where it was threatening the silk industry and the French economy. So when Napoleon became Emperor in 1804 he instructed that silk and not cotton would be worn as the ceremonial dress. The Romantic period at the turn of the 19th century saw the use of small floral prints. They were popular for their aesthetics and also because the small designs easily hid dirt spots and poor manufacturing.

In 1834 Perrotine printing was invented and used for the mass production of cloth. This process was the mechanisation of wood-block printing and allowed for multicoloured designs.

Polychrome patterns that had previously been produced through woven cloth could now be produced through a cheaper printing method.

In the 19th century lace manufacture was also mechanised. Large lace shawls made in the French towns of Valenciennes and Alençon became popular.

In the 1830s the jacquard was widely used. This was produced on a mechanised drawn loom and allowed for more complex weave structures and patterns.

It was felt by some in the late 19th century that technical advancements and mechanisation were responsible for a decline in the quality of design and crafts. Where a craftsperson had once been a designer and maker, the mechanised process was separating these two roles. The quality of textiles was poor and design was lacking. In Britain, William Morris was concerned with this situation and promoted handcrafted over machine manufacture. He designed textiles on naturalistic and medieval themes and chose not to use aniline dyes, preferring to dye them naturally.

He was the most prominent member of the Arts and Crafts Movement in England. Art nouveau developed from the Arts and Crafts Movement, with textiles becoming more stylised and intricately linear in design.

Opening Japan to international trade in 1854 resulted in the Japanese style coming to the West. Oriental motifs and Eastern flora, like the ayame pattern (a flower from the iris family) and also the chrysanthemum, began to feature in textile design.

Japanese lacquered products influenced the creation of shiny, lamé fabrics. In the 1860s, tarlatan, a thin plain, woven cotton, which was washed or printed with a starched glaze, was popular.

1900s

In the first quarter of the 20th century the Omega Workshops in London and Atelier Martine decorative art school and workshop in Paris opened. The Atelier Martine was founded by the couturier Paul Poiret, who was inspired by a visit to the Wiener Werkstätte school in Germany. The Atelier employed young girls with no design training who produced very naive textiles. This approach and look was in-line with the fauvist and cubist movements of the time in the fine arts.

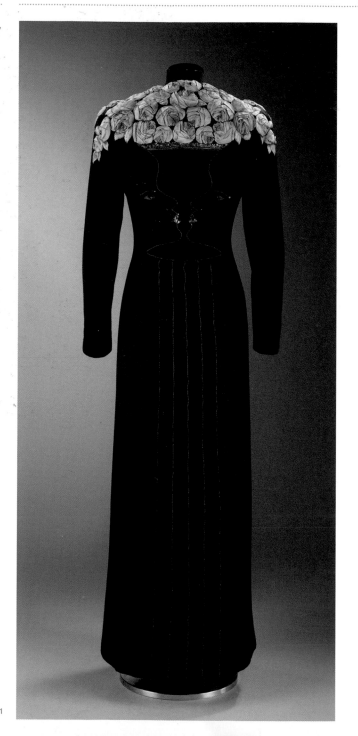

1920s

After the discovery of Tutankhamun's tomb in 1922 Egyptian motifs were translated into textile designs. The art deco style originated from the Exposition Internationale des Arts Décoratifs et Industriels Modernes exhibition in Paris in 1925. Looser shaped clothing became fashionable, influenced by the kimono shape and unstructured Eastern clothing. Madame Vionnet developed the bias cut, while Mariano Fortuny was inspired by classical clothing and created the pleated, unstructured Delphos dress.

During the roaring 1920s and the jazz era the new dance crazes called for dresses made from fabrics that moved on the body or seemed to under light. Fine, light fabrics, beading, sequins and fringing achieved this. Lace, fur and feathers were also popular for evening wear in this exciting and glamorous period. Viscose rayon was a popular fabric of the 1920s. This period also saw the introduction of the screen-printing process.

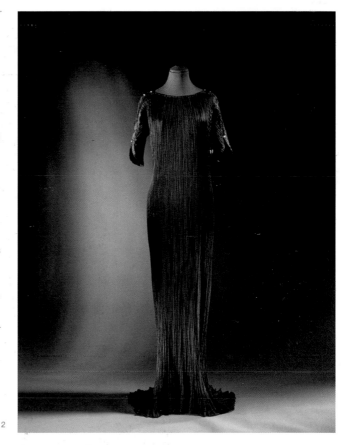

2

1930s

1   Full-length evening coat in black silk jersey, with appliqué pink silk flowers. Designed by Elsa Schiaparelli with Jean Cocteau; London, 1937.

2   'Delphos' evening dress in black pleated silk and decorated with Venetian glass beads. Designed by Mariano Fortuny; Venice, c.1920.

In the 1920s and 1930s Coco Chanel used jersey in day dresses. This was revolutionary, as this fabric had only been used before in underwear production. Florals, abstract and geometric patterns were popular, featuring two or more contrasting shades in a print. The development of cinema saw luxurious fabrics used for their lustre onscreen. Nylon was invented in 1935. Two-way stretch wovens were also developed.

Surrealism also influenced textiles. The first pullover Elsa Schiaparelli displayed in her windows created a sensation: it was knitted in black with a trompe l'oeil white bow. She was a close friend of the artists Salvador Dalí, Jean Cocteau and Christian Bérard and commissioned them to design textiles and embroidery motifs for her dresses. Schiaparelli experimented with unusual fabrics in her designs, including the modern fabrics rayon, vinyl and cellophane.

Researching textiles > Designing textiles

1940s

Fabric was rationed during the Second World War so the amount used within a garment was conserved, for example, skirts were slim, not flared or pleated, and were a shorter length. Jackets were single breasted and trousers were a specific length. This was the era of 'make do and mend' with people recycling their textiles. Dresses were made from curtains, clothes were altered and knitwear was unravelled and re-knitted. Silk supplies from Japan were cut off during the war, so nylon became a popular substitute. As France was occupied, Paris as a fashion capital was under threat and American fashions rose in popularity. Denim and gingham labourers' uniforms entered the ready-to-wear American market.

1950s

After the war there was a reaction against ornate pattern. Textiles featured futuristic imagery, scientific diagrams and bright, abstract shapes that echoed this atomic era. Textiles with linear drawings of newly designed domestic objects were also very fashionable.

With the end of rationing skirts became fuller and fuller. These circle skirts were often hand painted and embellished. The influence of America on Europe also saw Hawaiian shirts and American prints becoming increasingly popular.

Some of the couturiers, such as Balenciaga, created silhouettes that worked away from the body. They were interested in the space between the body and the garment. Stiffer fabrics worked well for this.

During the 1950s new fabrics were developed. These included:

Acrylic (1950)
Polyester (1953)
Spandex (1959)

1

1   Calyx furnishing fabric. Screen-printed linen, designed by Lucienne Day for the 1951 Festival of Britain. Manufactured by Heal's (1951).

2   Paco Rabanne mini-dress in perspex pailletes and metal chain.

The textile sample

1960s

Baby boomers reached their teens and wanted to be different from their parents, so they chose to wear shorter skirts and modern fashions. Textiles were zany, in bright colours. Space travel influenced bold prints and new synthetics with new dyes were being developed. Pierre Cardin and Paco Rabanne experimented with modern fabrications not seen in couture before.

Trousers were normal daily dress for women. Jeans also became very popular particularly amongst teenagers as a result of American westerns and the influence of movie stars such as James Dean.

Towards the end of the 1960s there was a nostalgic look back to the art deco and art nouveau periods. Imagery was enlarged and translated into bright psychedelic colours. Florals were depicted flatter and with bold colour, and the term 'flower power' was coined. The work of Finnish designer Marimekko illustrates this very well.

2

1

1970s

1980s

The unisex hippie folk movement was a reaction to the modernism and mass consumption of the 1960s and was triggered by the Vietnam War. Anti-establishment looked to different non-Western cultures and religions for inspiration and enlightenment. Fashionable men wore bright colours, lace and frills.

The oil crisis of the 1970s contributed to the downturn of the synthetic fibre market in Britain. Natural fabrics were increasingly adopted. In the UK Laura Ashley produced hand-printed looking cotton with Victorian florals.

The UK was politically and economically more stable and fashion followed suit, adopting a more conservative approach. In 1979 Margaret Thatcher became the first female prime minister of Great Britain. More women were working and they chose to wear tailored suits with large shoulders. The term 'power dressing' was coined. There was also a body-conscious trend with underwear worn as outerwear. Gaultier famously designed Madonna's conical bra outfits for her world tour in 1990.

Azzedine Alaïa and Bodymap designed with the developed stretch fabric Lycra to contour the body. There was also a different trend developing started by the Japanese designers Rei Kawakubo and Yohji Yamamoto. Garments were not body conscious, but played with interesting cut. Fabrics were monochrome, non-decorative and in some cases torn and raw. Recycled cotton was also introduced.

2

1990s

The trend started by the Japanese designers continued and was also taken up by a handful of Belgian designers. Martin Margiela was one of them; he worked in a conceptual way and wanted his clothes to look man-made not mass-produced. He used deconstruction and recycling throughout his collections. Ripped denim and customisation became mainstream.

2000s

Textiles have become more and more decorative as production is taken to the Far East and China. The factories here can add value to a textile through embellishment; the workers are skilled (often using local crafts) and the fabric can be produced cheaply. Modern fabrics are developing so that they are light-sensitive and breathable. Computer-aided design and manufacture is common. The designer is now far more in control of the mechanisation process, however, as a result, craft skills are unfortunately declining in Europe.

1   A 1970s textile design.

2   A range of dress patterns from the 1960s to the 1990s.

1

2

Different cultures

By looking at other cultures we can see the variety of uses for traditional textiles. In Japan the kimono is made from lengths of fine woven silks and there is little cutting in manufacture so that the pattern of the cloth can be clearly seen. This is in contrast to the Western tailoring of the 16th and 17th centuries. A garment that had seaming, darting and panelling was very desirable, as it would have been expensive to produce and would indicate that the wearer was wealthy enough to afford such a garment.

It is interesting to look to other countries and their traditional handcrafted textiles for inspiration and to note how these techniques can be applied to modern textiles.

1   Indian textile designs.

2   Hand-crafting textiles in Cambodia.

3–4   Examples of research boards.

3

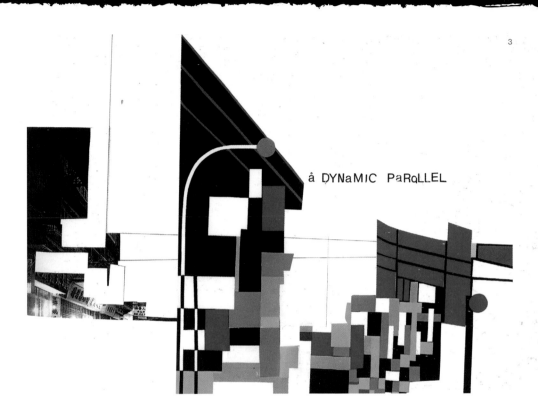

à DYNaMIC PaRqLLEL

Non-classic inspiration

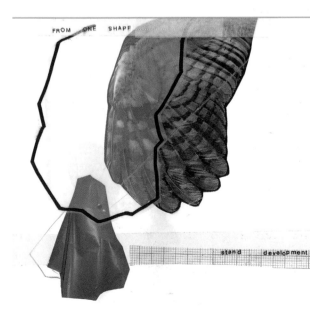

Some of the most experimental
textiles come from not looking at
existing textiles, but from looking
at something completely different
for inspiration, such as nature or
architecture, the fine arts or
contemporary culture.

4

Designing textiles

1   Example of a mood board.

2–3   Examples of drawing
techniques.

The next step is to collate the research that you have gathered.
This gathering of informative textiles from classic textiles, cultures
or other sources of inspiration can be in the form of mood boards or
sketchbooks that document the research and create links to form a
story that will develop into design ideas. It is important to then find
a way to express your design ideas through drawing, collage,
photography, or maybe CAD work. It is wise to also consider the
surface you are going to design on: will you start to work on paper
initially and then develop into cloth and knit, or will you start to work
directly with material? As you design you must understand the
basic textile design principles of scale, texture, colour, pattern,
repeat, placement and weight. Consider how these principles work
within a sample and how these samples work together as ranges,
as well as how your designs will result in functional, inspirational or
commercial textiles suitable for use within contemporary fashion
design and garment construction.

Rendering designs

You now need to think about what
you are trying to design and how
best to go about it. Determining the
most appropriate medium to render
your designs in is very important,
whether it's paper, paint, pencils or
a software package. Work out what
is required and in what time frame.
Bear in mind that you might need to
learn new skills for the designs you
are creating. Always remember to
experiment and enjoy the process.

2

Drawing

Being able to communicate your
ideas through drawing is fundamental
to most design disciplines. However,
it is possible to also use other media
such as collage and photography as
a means of communication.
Experiment with drawing, use
different types of media and be
expressive with line, colour and
texture. Think also about silhouettes
and blocks of colour or tone within
your design. Consider whether you
are trying to represent what you are
drawing precisely or if the artwork is
developing in a more abstract
direction.

Researching textiles > **Designing textiles** > Textiles into production

3

Collage and 3D rendering

Working with different types of papers and building up layers to create textures can be useful for knit and weave ideas. Try finding unusual textures to play with, but remember to refer back to the function of your fabric. You might try to experiment and mock up a sample in a fabrication similar to the yarn you might eventually use.

1   Example of collage work.

2   Tata-Naka A/W07 collection featuring digitally printed textiles produced using CAD.

1

2

CAD

This stands for computer-aided design. The use of the computer can make the design process faster. Colour and scale can be changed more quickly than manually recolouring or rescaling a design. Remember that colours on a computer screen are different from those eventually printed out, as the computer screen works with light and not pigment. Scanning in original drawings and combining them with other imagery can work well. Avoid using filters and treatments from design packages unless they are used originally otherwise they can look very obvious.

Photography

The use of photography can be great for capturing ideas quickly. Textures and shapes can be registered in great detail immediately without the need for hours of drawing. With the use of packages such as Photoshop, images can now be successfully translated into designs. Layers and collages can be built up on screen.

1

Basic textile design principles

It is important that as a designer
you understand the basic design
principles of textile design. This
knowledge will allow you to fully
explore the design process.
Obviously different samples will
feature certain principles more than
others. For example, you might
produce a range of black samples
that focus on the application of
shiny surfaces to matt-base cloths.
The juxtaposition of surfaces and
placement of pattern might be the
focal point of these designs rather
than colour.

Scale

Look at the scale of your design
within the fabric piece. Is it very
small and repeated or is it enlarged
and abstract? You may consider
placing a large design with a smaller
design for added contrast. Think
about how this design will work on
the body and how it will work within
the pattern pieces of a garment. An
enlarged bold design may not have
as much impact if the design has to
be cut up to be used in a garment
with many pattern pieces. Think
how you can place a large design
within a garment silhouette for the
best effect.

1   Example demonstrating
    how print scale can work
    on the body.

2–3 Liberty print designs by
    Duncan Cheetham showing
    an all-over floral pattern
    (top) and a chevron print
    (bottom). The chevron
    design has a 'direction',
    a clear top and bottom
    to the design.

Pattern and repeat

If you would like your textile sample to work down a length of fabric you must consider how it repeats. Repeats can be very simple or very complicated working across a large area. The bigger the repeat the harder it is to see on a length of fabric; a small repeat is more obvious. It is important to observe how your design flows across a length. When you repeat your design en masse you might find that you can see where you are clearly repeating the motif. This might work in a design or it might look rather crude.

Also consider if there is a direction to your design. Is there a top and a bottom? This can look very interesting visually, but remember that this kind of design limits the lie of a fabric, as the pattern pieces will all have to be placed in one direction.

If you are working on a computer it is very easy to see how your design will work by cutting and pasting. There are also computer packages that quickly put your design into repeat. To work out manually whether your designs flow, cut the design in half and place the top part below the bottom to see where you need to fill in gaps.

2

3

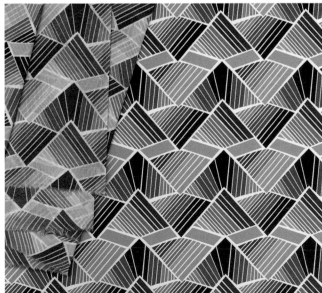

Placements and engineered designs

Placements work well if you consider the position of the design on the garment. The most obvious placement is a print placed on the front of a t-shirt. It is interesting to consider how a design can be engineered to work around a garment. Can a seam be moved to allow a design to travel from the front to the back of a garment? Could a placement work around the neck or around an armhole? Can a design fit into a specific pattern piece? If you are working in this way you may have to consider how the engineered design scales up or down according to the size of the garment. A size 10 garment will have a smaller neck hole than a size 14. You will have to produce a different size design for each dress size for this to really work. If you are working on the computer this is much easier as designs can be scaled quickly and placed within pattern pieces.

Clever use of placements might affect the construction of the final garment. For example, a coloured block could be knitted directly into a garment, which would mean a coloured panel would not need to be cut and sewn in. A weave could incorporate an area of elastic running across it, thereby avoiding darting in the final garment to fit it to the body. Smocking applied to a fabric can work in a similar way.

Colour and colourways

It is often a good idea to start finding a colour palette that you like and that suits your theme before you begin designing. Finding an image, a photograph or painting where the colours already work together can be a good start or you may just start selecting colours and working them together by eye. You can work with chips of paper colour, fabric swatches or on the computer. A palette of colours can be any size, but do not over complicate it by using too many colours. Check your balance of colour and tone within the palette. Consider what the colour is going to be used for and in what proportion. Remember a small area of colour looks very different to an expanse of the same colour over a couple of metres of fabric.

When you design consider the various tones and saturations that can be found within one colour. Also experiment with the different textures of a hue. For example, the colour black can be blue-black, warm black, washed-out black, matt black, shiny black, or transparent black. Your palette will change under different lighting conditions – natural light at certain times of the day and different forms of electric lighting will all have an effect.

1   Givenchy S/S08. This design contains a striking use of placement. The circles on the jacket are placed so they correspond to the circles found on the blouse and shorts beneath. The circles on the front of the jacket also align with those on the sleeves and cuffs. Catwalking.com.

2–4   Prints by Jenny Udale showing a matt print on a shiny fabric. Puff adds surface interest and colours work together.

Weight, texture and surface

When you start to transfer your designs on to or into fabric, think about what weight your textile will be in relation to the design and also in relation to its use in the final garment. Understanding fabrics and yarns is paramount to this process (this will be explored more later on in the book).

Consider whether your design would benefit from texture. Surface interest is very important within textile design, especially in knit, embroidery and embellishment. In knit and weave design the weight of the yarn and size and type of stitch or weave will affect the texture. For printed textiles, surface interest is achieved through printing. Some printing media will sit on top of the fabric and produce a relief effect, while others might eat away at the surface of the textile through a chemical reaction.

The type of embellishment and the yarn or stitch used will produce various textures on embroidered fabrics. Mechanical and chemical finishing processes can change the texture of a fabric after it has been created.

Interesting textiles can be created by experimenting with a mixture of processes, for example, pleating a fabric before you print on to it, or knitting a fabric then boiling it to give a matted texture.

Textiles into production

As a student you will be creating small textile samples and developing experimental and exciting ideas. You will probably only have to produce a small length of fabric or a small range of garments that feature your fabrics. However, when you become a designer in the fashion industry you will have to consider how you sell your work. If you choose to manufacture your textiles you will also have to consider the skills and technology you will need for production and the ethical choices you might make. You must consider how your textiles now work together and form a collection; then to whom you will present the samples and where you will sell them.

Collections of fabrics

When you create a collection of fabrics you must consider how the designs work together and what their common theme is. Are you creating a collection of similar designs, for example, a range of striped textiles or a variety of designs – a stripe, spot and floral – that are maybe all rendered by a similar drawing technique? Consider how your range of designs works within a fashion collection: do you have all the different weights and qualities needed for all the garments?

The colour palette is usually common to a range of fabrics, but you can vary the proportion of colour used in each sample within the range. Try hard not to repeat a motif in a collection of designs. For example, you might think each design is very different, that in one design your motif of, say, a leaf is small and lime green and in the next design it is larger and black, but one company may buy the first design and another the second, and their designers could then resize and recolour your designs and end up with similar textile designs.

1   A selection of fabrics used for Wildlifeworks' S/S08 collection (see page 43). The organic fabrics are digitally printed and include a variety of textures, weights, embroidery and embellishment.

Presentation

Calendar of Trade Fairs

**Paris, France
February/September**

Première Vision: promotes fabric for clothing.

Expofil: yarns and fibres.

Indigo: textile design including print, knit, embroidery and vintage fabrics.

Le cuir à Paris: leather, fur and textiles for accessories.

**New York, USA
January/July**

Première Vision: preview.

**Milan, Italy
February/September**

Ideabiella: menswear and womenswear fabric collections.

Ideacomo: fabrics for womenswear.

Moda In: avant-garde materials for the fashion market.

Prato Expo: fabrics for womenswear with a high fashion content and casual menswear.

Shirt Avenue: traditional and novelty shirting fabrics.

**Florence, Italy
January/July**

Pitti Filati: yarn show.

Textile samples tend to be presented on hangers or simply mounted on light card fixed at the back. It is important that the textile is not stuck down as it needs to be handled, therefore usually only one edge is attached to the mount leaving the fabric sample hanging so the weight and drape can be experienced. Keep the mounting plain and simple so it does not distract attention away from the textile design. It is not normally advisable to present your samples in portfolio plastic sleeves, as the fabrics cannot be easily handled.

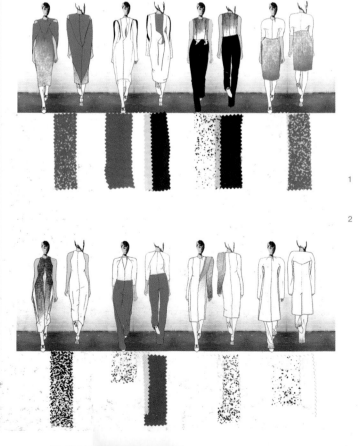

1

2

Fabric fairs

1–2 Fabric swatches presented alongside fashion drawings clearly show how the textiles will be used within a collection. The fabrics are not fully stuck down, but hang so they can be handled easily.

3 The Première Vision fabric trade fair.

Fabric trade fairs are held biannually in line with the fashion calendar. The fairs showcase new developments in woven, knitted, printed and embellished fabrics.

Première Vision is the main fabric and colour fair held biannually in Paris. Fabric manufacturers from around the world display their new fabric samples and take orders from designers. Sample lengths of fabrics are made first by the manufacturer and sent out to the designer. From this, garment samples are made and orders are taken. Based on this, the fabric is ordered and if not enough is ordered by designers, a fabric will not go into production.

Indigo, also held in Paris, is a platform for textile designers (mainly print designers) to show their textile samples. The samples are shown as collections and are bought by designers for inspiration or by fabric companies and fashion companies to be put into production.

Pitti Filati is a biannual yarn fair held in Florence. Here yarn companies display their latest collections of yarns for production and textile designers sell their knitted and woven samples. The other main yarn fair is Expofil in Paris.

If you choose to represent yourself at a fabric fair you must consider the cost of travel, hiring a stand at the exhibition, manning the stand and accommodation while you are there. If an agent takes your work to sell they will take a large cut of the sales of your samples to cover their expenses. Always keep a good record of the samples that you give to an agent. Number each sample on the back and list the ones that are going, get the agent to confirm and sign the list. Make sure you know what percentage the agent is taking and how long they will take to pay you.

3

Future fabrics

Developments in the creation of textiles seem to be following two paths – ethically driven by the environment and future-technologies driven by scientific advances – and where they meet is where great future fabrics will be produced. In other words, fabrics that use great design and can be sustainable, but can also be forward thinking.

We should also consider how traditional crafts, such as block printing, hand crochet and crewel work can be maintained. These handcrafts give textiles character and individuality, and they can add value to a product as a result of the time and skill needed to create it. A garment that has been hand stitched and embroidered will never be exactly the same as another garment. Certainly high-end designers are incorporating handcrafted fabrics and finishes into their collections, but these handcraft techniques are difficult for the high street to copy and therefore set them apart. Consumers, however, are demanding fabrics that can perform well and that can wash and wear well, so maybe combining craft with performance and modern technologies will ensure their survival.

1   A design from Sandra
    Backlund's Ink Blot Test
    collection.

2   An embellished design
    from Alabama Chanin's
    S/S08 collection.

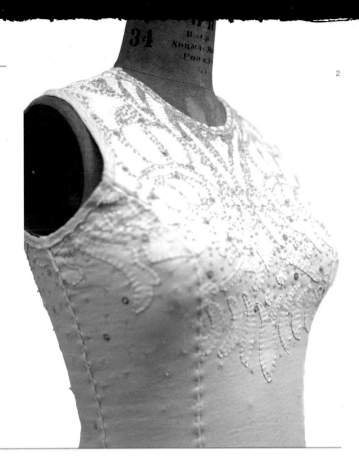

2

Ethical

Clothing is becoming cheaper and cheaper as production is getting larger and larger. We are buying our clothing in supermarkets with our weekly food shop. We are wearing a t-shirt a few times and throwing it away to buy the next desirable cheap garment. Fashion has a short shelf life with new collections appearing every six months. If the season's collections do not sell in the season they go on sale, they are burnt or recycled.

One reaction to this mass consumption is the rise of sustainable collections. Companies are considering what the impact of their textiles and processes has on the environment. Many are choosing to use fabrics that are made from recycled material, either at fibre or fabric level. Many fibres come from natural sources and can be reused; some synthetic fibres can also be recycled, for example, polyester can be made from old plastic bottles. Dye companies

that use synthetic dyes are reducing the amount of chemicals that are needed in processes and recycling the water they use, so reducing the impact of production. Synthetic dyeing is often seen as unethical. However, natural dyes need fixers that can be harmful to the environment as they build up; also some natural dyes need a large amount of natural material to produce a small amount of dye.

There has definitely been a trend for organic and fair trade in industries such as food and cosmetics, but the fashion industry has been slower to pick up on the idea. Some may say that fashion is fundamentally about aesthetics, so is there room in fashion for ethics? It is important that ethical companies integrate functionality, design and quality into their ethical story for their products to be fashionable and desirable. They will, however, be competing with low-price manufacturers who

are churning out products more cheaply and quickly than before.

As a designer you can choose where you buy your textiles or where you have your textiles manufactured. It may be harder to source sustainable or ethical material and it may make your designs more expensive. You may be competing with cheaper goods from non-certified factories, but ultimately it is your choice. Decide how much you want to be involved with the issues, but educate yourself.

Fairtrade

The term 'fairtrade' is part of the Fairtrade Foundation's logo and is used to refer to products that have actually been certified 'fairtrade'. The Fairtrade Foundation gives this certification after it checks that the growers or workers have been given fair pay and treatment for their contribution to the making of the product.

The working environment in which the products are made is taken into account. Manufacturers have to demonstrate that they provide good conditions for the people involved in the factory. There are basic standards covering workers' pay and conditions, as well as issues such as the absolute prohibition of the use of child labour, which must be met in order to qualify for the fairtrade 'kite mark'.

Fairtrade is also used to describe products that try to encourage the use of natural and sustainable materials, together with contemporary design to maintain ancient skills and traditional crafts, where regular employment and the development of skills can bring dignity back to people and their communities.

1

Organic

The General Assembly of the International Federation of Organic Agriculture Movements (IFOAM) is the worldwide umbrella organisation of the organic movement, uniting 771 member organisations in 108 countries. IFOAM's goal is the worldwide adoption of ecologically, socially and economically sound systems that are based on the Principles of Organic Agriculture. The principles aim to protect the land that is being farmed and also those working on it and the communities of which they are a part.

Strict regulations define what organic farmers can and can't do, placing strong emphasis on protecting the environment. They use crop rotation to make the soil more fertile. They can't grow genetically modified crops and can only use – as a last resort – seven of the hundreds of pesticides available to farmers (see Chapter Two: Fibres, for more information on organic cotton production).

Animal rights

The campaign for animal rights gets stronger every year, yet designers continue to show catwalk collections that contain fur. There still seems to be a demand by a certain consumer group for fur in fashion. Designers are now using fur and leather substitutes in experimental ways. Stella McCartney does not use any animal products in her collections; instead she uses canvas and pleather (fake leather) in her accessories. There is a lot of research to develop good leather-look fabrics. The Japanese company Kuraray produces Clarino and Sofrina and the company Kolon Fibers produces an ultra-microfibre textile called Rojel.

Technology

Technology is being used to generate new fabrics and also to produce existing fabrics more quickly and efficiently. The possibilities of futuristic textiles are positively endless.

Smart materials

Interactive clothing incorporates smart materials that respond to changes in the environment or to the human body. Heat, light, pressure, magnetic forces, electricity or heart rate may cause changes to shape, colour, sound or size. It is especially appropriate to textiles, as during the construction process, fibres and yarns can form circuits and communication networks through which information is transferred. Coating finishes, printing and embroidery can also all be used to conduct information. Clothes could quite possibly interact directly with the environment by opening doors or switching on lights or could communicate with images, light or noise.

2

Biotechnology

Fabrics can contain chemicals within their fibres that can be released on to the skin for medicinal or cosmetic reasons. Fibres are being developed from natural sources to mimic nature, for example, the development of spider silk. Fabrics are also being grown directly from fibres in the same way that skin or bones grow.

CAD

Digital technology and computer-aided design is advancing and making the designer's job easier. Designing a textile sample using CAD can produce a repeat in many colours far quicker than if done by hand. Computerised looms can produce metres of fabric in minutes. Obviously manufacturing processes must evolve, but it is important to still understand the craft techniques on which these processes are based.

1   A design from Wildlifeworks A/W07 collection. Wildlifeworks produces organic and fairtrade clothing.

2   This high-tech dress from Hussein Chalayan's A/W07 collection features a light display system within its structure so that the patterns within the textile can be changed.

'As a child I was obsessed with video games, now I am fascinated with technology and what you can do with it.'

Rory Crichton, freelance printed textile designer for Giles Deacon and Luella Bartley

1   Vintage Commes des Garçons suit. The jacket is 100% polyester with cotton panels and the skirt is 100% polyester with velvet ribbon, lace and tape work.

Fabrics are made fundamentally from fibres. These fibres can be categorised simply as natural or synthetic and each fibre has its own characteristics and qualities. For example, cotton fibres produce a fabric that is breathable, while wool fibres create a warm cloth, but one that can be sensitive to heat. The way the fibres are spun and the yarn constructed affects the performance and look of the final fabric. Finishes and treatments can be applied to a textile at fibre, yarn, cloth or final garment stages of production. These finishes can enhance and change the qualities of the textile for fashion. Colour, texture and performance qualities can all be added. Obviously the way the fabric is constructed also gives the fabric a specific quality. This will be discussed in the next chapter.

Companies who manufacture man-made and natural fabrics are considering their impact on the environment with their manufacturing processes. The production of natural fabric may have more impact on the environment than a man-made one if it uses harmful chemicals in its processes; also many man-made fabrics can now be completely recycled. Fabric characteristics can be integrated into the make-up of man-made fibres reducing the need for chemical and mechanical finishing processes.

Natural

1   Top row (from left to right): silk organza, silk jersey, raw silk, dupion silk; wool herringbone, wool melton wool; shearling, leather, horse hair.

Bottom row (from left to right): foil-printed linen, linen; denim, cotton shirting; bamboo, jute hessian.

2   A Jessica Ogden cotton voile with a spot weave and nylon net dress.

3   Wildlifeworks S/S08 collection featuring organic cottons.

Natural fibres are derived from organic sources. These can be divided into plant sources (composed of cellulose), or animal sources, which are composed of protein.

1

Cellulose

Cellulose is made of carbohydrate and forms the main part of plant cell walls. It can be extracted from a variety of plant forms to make fibres suitable for textile production. Here we are looking at fabrics that are most suitable for the production of garments; they must be soft enough to wear and not break up when worn or washed.

2

Cotton

Cotton is a prime example of a plant fibre. It has soft, 'fluffy' characteristics and grows around the seed of the cotton plant. These fibres are harvested from the plant, processed and then spun into cotton yarn.

Cotton fibres are used to produce 40 per cent of the world's textiles. Its enduring popularity is its extreme versatility; it can be woven or knitted into a variety of weights. It is durable and has breathable properties, which is useful in hot climates as it absorbs moisture and dries off easily. The longer the fibre, the stronger and better quality the fabric is, for example, Egyptian cotton.

Cotton is mainly produced in the USA, China, the former Soviet Union, India, Mexico, Brazil, Peru, Egypt and Turkey. In most cotton production, farmers use chemical fertilisers and pesticides on the soil and spray them on the plants in order to prevent disease, to improve the soil and to increase their harvest. Cotton has always been extremely prone to insect attack and since insects started building up immunity to pesticides, the situation has worsened. This means growers have increased their use of chemical pesticides simply to ensure crop survival. Cotton crops in India, America and China demand thousands of tonnes of pesticides, which are sprayed on fields from the air. This overuse of pesticides is rendering hundreds of acres of land infertile and contaminating drinking water. The World Health Organisation estimates that about 20,000 people die each year as a result of pesticide use.

Also the chemicals that are used are absorbed by the cotton plant and remain in the cotton during manufacture, which means that it is still in the fabric that we wear next to our skin. Due to these issues, manufacturers are increasingly developing organic fibres that are grown and processed without the use of artificial fertilisers and pesticides. Organic fabric production is more expensive, but it has a low impact on the environment and is healthier for the consumer. There are designers pursuing organic solutions such as Katharine Hamnett, Wildlifeworks and Edun.

Linen

Linen has similar properties to cotton, especially in the way it handles, although it tends to crease more easily. Linen has good absorbency and washes well. It is produced from the flax plant and is commonly regarded as the most ancient fibre.

Hemp, ramie and sisal are also used to produce fabrics as an alternative to cotton.

Natural > Man-made

Protein

Protein is essential to the structure and function of all living cells. The protein fibre keratin comes from hair fibres and is most commonly used in textile production.

Wool: cashmere, angora and mohair

Sheep produce wool fleece for protection against the elements and this can be shorn at certain times of the year and spun into wool yarn. Different breeds of sheep produce different qualities of yarn. Merino sheep produce the finest and most valuable wool. 80 per cent of wool is produced in Australia, New Zealand, South Africa and Uruguay. Biodegradable and non-toxic pesticides are now more widely used in the production of wool to protect the sheep and improve the environment.

Goats are also used to produce wool; certain breeds produce cashmere and angora. Cashmere is extremely soft and drapes well. Alpaca, camel and rabbit are also sources of fabrics with a warm, luxurious feel to them. Wool has a warm, slightly elastic quality, but it doesn't react well to excessive temperatures; when washed in hot water it shrinks due to the shortening of the fibres.

1   Christian Wijnants A/W07. Heavy hand-knitted jumper made from wool and angora.

2   Dolce & Gabbana A/W07. Ostrich feathers trapped beneath silk voile. Catwalking.com.

1

Silk

Silk is derived from a protein fibre and is harvested from the cocoon of the silkworm. The cocoon is made from a continuous thread that is produced by the silkworm to wrap around itself for protection. Cultivated silk is stronger and has a finer appearance than silk harvested in the wild. During the production of cultivated silk the larva is killed, enabling the worker to collect the silk and unravel it in a continuous thread. Silk worms live off mulberry trees. For one kilogram of silk, 200 grams of leaves must be eaten by the larva. Once extracted from the cocoon, the larva is often used as fish food by the farming community. In the wild, the silkworm chews its way out of its cocoon, thereby cutting into what would otherwise be a continuous thread. Silk fabric has good drape, handle and lustre.

2

Fur

Animals such as mink, fox and finn raccoon are bred on farms where the animals are purely reared for their skin. This subject causes heated debate between those for and against fur. The fur farmers would argue that the ethical treatment of the animals has always been an important part of the approach to fur farming. The quality of fur depends on the welfare of the animal; the higher the quality of life the better the quality of fur. Fur farmers in Scandinavia are regulated by national laws and guidelines, and regulations governed by the Council of Europe. Scientists work closely with the farmers and their research findings have already been adopted in areas such as housing, disease prevention, nutrition, husbandry, breeding and selection. The process of implementing other animal welfare measures as a result of scientific research is ongoing.

Anti-fur protesters would argue that some fur used in the clothing industry is used from animals 'caught' in the wild. The animals are trapped in snares or traps and undergo hours of suffering before they are brutally killed. As for 'farmed' fur, again they will argue that the millions of animals killed every year are kept in small, cramped cages or enclosures. These living conditions go against the animal's natural instincts and cause severe stress and can lead to cannibalism and self-mutilation. The killing process itself is not quick and painless. Methods including gassing, poisoning, electrocution, suffocation and neck breaking are commonplace.

Leather

Leather is made from animal skins or hides. The procedure used to treat the raw animal hides is called 'tanning'. First the skins or hides are cured, a process that involves salting and drying, then they are soaked in water. This can be for a few hours or a few days. The water helps to rid the skin of the salt from the curing process as well as dirt, debris and excess animal fats. Once the skins are free from hair, fat and debris they are de-limed in a vat of acid. Next the hides are treated with enzymes that smooth the grain and help to make them soft and flexible. The hides are now ready for the 'tanning' process. There are two ways of tanning – vegetable tanning and mineral

tanning. Which type is used depends on the hide itself and the product intended. Vegetable tanning produces flexible, but stiff leathers that are used in luggage, furniture, belts, hats and harnesses. Mineral or chrome tanning is used on skins that will be used for softer leather products, such as purses, bags, shoes, gloves, jackets and sandals. The skins then go through dyeing and rolling processes, which dry and firm the leather. The final step of the process involves finishing the skin. This is done by covering the grain surface with a chemical compound and then brushing it. Some leathers will show lots of imperfections after their final finishing, but they can be

buffed or sandpapered to cover these up; after a period of prolonged buffing the leather becomes suede. 'Splitting' the leather can also produce suede whereby the skin is cut into layers or splits with the outer or top layer being leather and then all the lower layers being suede. The higher-quality suede is in the upper layers.

Leather stretches but does not return to its original shape. DuPont has developed a fabric that is a fusion of leather and Lycra that has the properties of both. Napa is soft thin leather used for garments and can be made from leather skins or suede.

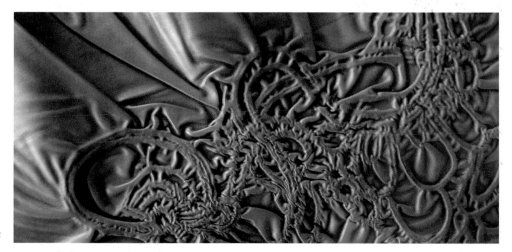

2

Metal

Hides and skins

Hides come from large animals such as cows, horses and buffalo, while skins are from smaller animals such as calves, sheep, goats and pigs.

Fibres can be drawn from metal rods; alternatively metal sheets can be cut into very fine strips. Metallic fibres can be used to decorate clothing; traditionally gold or silver strips were used but they are fragile and expensive, and silver tends to tarnish. Nowadays aluminium, steel, iron, nickel and cobalt-based superalloys are used.

1   Kenzo cow-hide skirt.

2   Sophie Copage leather design. The pattern is achieved through the application of heat.

Man-made

1   Man-made fabrics. Top row
     (left to right): spun rayon,
     tencel/lyocel, Nylon/
     elastane, Nylon ripstop.

     Middle row (left to right):
     silk/viscose/spandex
     /velour, nylon fusing,
     viscose, satin viscose.

     Bottom row (left to right):
     polyester, polyester
     wadding, metallic silk.

2   Rory Crichton's *Strange I've
     seen that Face Before*
     textile design.

Man-made fibres are made from cellulosic and non-cellulosic fibres. Cellulose is extracted from plants, as well as trees. Man-made fibres such as rayon, Tencel, acetate, triacetate and Lyocell are cellulosic fibres as they contain natural cellulose. All other man-made fibres are non-cellulosic, which means they are made entirely from chemicals and are commonly known as synthetics.

Developments in the chemical industry in the 20th century caused a radical transformation in fabric production. Chemicals that had previously been used for textile finishing techniques began to be used to extract fibres from natural sources in order to make new fibres.

Cellulosic fibres

These are fibres that are derived from cellulose, but through chemical manufacturing processes are developed into new fibres.

Rayon

Rayon was one of the first man-made fabrics to be developed. The first rayon dates back to 1885 and was called artificial silk, due to its properties. The name rayon was not established until 1924. As it is derived from cellulose (wood pulp) it has similar qualities to cotton in that it is strong, drapes well and has a soft handle. Rayon has excellent absorbency, so it is comfortable to wear and dyes well. Different chemicals and processes are used in the production of rayon, each with its own name. These include acetate rayon, cuprammonium rayon and viscose rayon, known commonly as viscose. Lyocell and Modal are evolved from rayon.

2

Cellulose acetate

Cellulose acetate, more commonly known as acetate, was introduced during the First World War as a coating for aeroplane wings and was then developed into a fibre. It is made from wood pulp or cotton linters. Acetate shrinks with high heat and is thermoplastic, and it can be heat set with surface patterns such as moiré. It has the look, but not the handle, of silk. It does not absorb moisture well, but is fast to dry.

Tencel

Tencel was more recently developed to be the first environmentally friendly man-made fabric. It is made from sustainable wood plantations and the solvent used to extract it can be recycled, so the Tencel fibre is fully biodegradable. It produces a strong fabric that drapes like silk, with a soft handle.

Non-cellulosic or synthetic fibres

Germany was the centre of the chemical industry until after the First World War when the USA took over its chemical patents and developed its inventions. DuPont was one of the large chemical companies developing fabrics at this time. In 1939, DuPont was able to produce long polymeric chains of molecules, the first being the polymer nylon. This was the beginning of the development of synthetic fabrics.

Most synthetics have similar properties. They are not particularly breathable, so many are not as comfortable to wear as natural fibres. They are sensitive to heat, so pleats and creases can be set permanently, also fabrics can be glazed or embossed permanently. However, unwanted shrinkage and glazing can occur when a finished garment is pressed.

In general synthetic fibres are white unless they are first dyed. Synthetic fabrics have poor absorbency, which means they dry quickly, but it makes them difficult to dye. Dyeing at the fibre stage of production produces a very colourfast fabric, but it means that the fabrics produced in this way cannot respond quickly to fashion trends, as the colour is determined early on in production.

Nylon

Nylon is a strong, lightweight fibre, but it melts easily at high temperatures. It is also a smooth fibre, which means dirt cannot cling easily to its surface. It has very low absorbency so dries quickly and doesn't need ironing. Nylon is made from non-renewable resources and is non-biodegradable. During the Second World War silk supplies from Japan were cut off, so the US government redirected the use of nylon in the manufacture of hosiery and lingerie to parachutes and tents for the military.

Lycra is a form of nylon and was developed to use in lingerie, sportswear and swimwear.

Acrylic

DuPont developed acrylic in the 1940s. It has the look and handle of wool, but pilling can be a problem. It is non-allergenic, easy to wash, but sensitive to heat and melts under high temperatures.

1

1   Jean-Pierre Braganza A/W07. Mercury lurex jersey dress with leather collar.

2   Junya Watanabe dress made from 100% polyester. The dress is made from a continuous piece of fabric that wraps around the body creating a semi-opaque finished garment.

Polyester

Polyester is a strong, crease-resistant fibre developed in 1941 by ICI. It is the most widely used synthetic fibre and is most commonly found in blends where it is used to reduce creasing, softening the handle of the cloth and adding drip-dry properties. Polyester was introduced to the USA as Dacron.

Polyester is made from chemicals extracted from crude oil or natural gas by non-renewable resources and the production of fibres uses large amounts of water for cooling. However, polyester can be seen as an environmentally friendly man-made fabric; if it is not blended it can be melted down and recycled. It can also be made from recycled plastic drinks bottles.

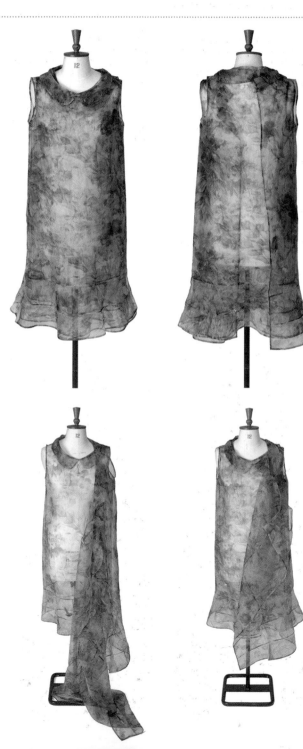

2

Spandex

Spandex is a super-stretch fibre as it can be stretched 100 per cent and will return to its original length. It was introduced by DuPont in 1959 and is a manufactured elastic fibre; it has similar properties to natural rubber, which is a natural elastic fibre. Spandex is used to add power stretch or comfort to textile products. Power stretch provides garments with holding power and is often used in underwear or swimsuits whereas comfort stretch adds only elasticity.

Aramid fibres

DuPont introduced aramid fibres under the trade name Nomex nylon in 1963. The fibre is also known as Kevlar. The fibres have exceptional strength and are five times stronger than steel. They are also flame resistant, decomposing at temperatures of about 371°C (700°F). Kevlar is used for strength applications where the fabric needs to be light, for example, in bulletproof outfits. Nomex is used for its flame-resistance properties especially in military clothing and firefighting uniforms.

1   Rubber strips.

2   Shoe design by Marloes ten Bhömer incorporating a carbon fibre structure. The heels are positioned at the side of the shoe, making the wearer walk slightly differently, continuously moving the weight of the body from side to side.

New fibre developments

Chemists are now producing fibres from natural sources, changing their structure to produce superior properties. They are also developing microfibres and nanotechnology, which can produce fabrics with advanced properties that can react to the environment in various ways.

1

Odin Optim

Is a fibre developed by Nippon Keori Kaisha, Ltd in Japan, The Woolmark Company and the Commonwealth Scientific and Industrial Research Organisation. They have taken the wool fibre and altered its structure to produce a wool fabric that has superb drape and tactile qualities.

Azlon

This is a generic name for fibres regenerated from milk, peanut, corn and soybean proteins. Japan has produced a fibre made of milk protein and acrylic called Chinon. It resembles silk and is used for garments.

Spider silk or BioSteel

Spider silk is naturally stronger than steel and is stretchy and waterproof. Biochemists are currently studying its structure and developing synthesised fibres with the same properties that could be used for fabric production. It is derived from protein in goat's milk and is trademarked BioSteel.

PLA fibre

This is a fibre that started being developed in 2001 under the trade name NatureWorks. It is derived from naturally occurring sugars in corn and sugar beet. The fibre is produced from a renewable source, needs little energy in production and is recyclable.

Microfibres

Microfibres are extremely fine fibres of one denier or less that have advanced properties. These fibres can be engineered into the construction of a fabric or can be used as a finishing coating. Microfibre properties may include being lightweight, tactile, water resistant, windproof or breathable and they are often used in sportswear and high-performance clothing. As their properties are integral to the fibre they will not wear or wash off. The microfibre Tactel is produced by DuPont and has great tactile properties; there are different types of Tactel all with their own advanced properties. Microfibres can be more expensive to manufacture so they are often mixed with cheaper fibres.

Microfibres can be produced with microcapsules that contain chemicals such as medication, vitamins, moisturisers, antibacterial agents, UV blockers or perfume. Chemicals in the microcapsules can be released on to the skin either by abrasion or as a result of heat given off by the body. Medication, vitamins or moisturisers can then be absorbed, imparting their benefits to the skin. However, these chemicals do get used up and gradually wash out of the fabric. Micro-organisms can also be incorporated that live off dirt and sweat, therefore maintaining cleaner odour-free garments.

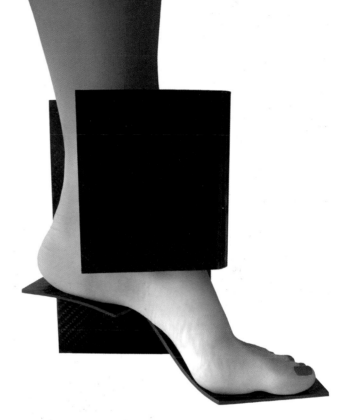

2

X-static

Metals can be woven into fabrics to make them more malleable and they are often mixed with synthetics for their anti-static properties. The use of silver is being developed within fabrics as a result of its antibacterial properties. X-static is produced by Noble Fiber Technologies and bonds silver to the surface of another fibre to give it advanced properties.

Nanotechnology

Nanotechnology works at a molecular level, ultimately creating extremely intelligent and sophisticated fabrics that could be used in garments to change their colour, structure and even size. Currently nanotechnology is used for producing finishings for fabric, for example, Schoeller has developed a dirt-resistant coating for fabrics.

Natural > **Man-made** > Yarn

Yarn

1   Shown anticlockwise (from top right): raw, unfinished sheep's wool fleece; spun wool; raw cotton (cleaned, but as grown from the plant); raw cotton (cleaned), carded and straightened; spun cotton yarn; elastane/lycra yarn; raw viscose rayon fibre (which is wood pulp dissolved and regenerated as viscose fibre); polyester yarn polymer chips melted and extruded as a continuous filament; linen yarn from flax plant; raw silk yarn spun from silk cocoons.

2   Shown from left to right: linen loop; wool loop; crêpe; tape; mohair; linen; raw silk; silk; cotton slub; wool slub; chenille.

Most fibres go through a process that produces a yarn, which then goes through a construction process to produce a fabric. (Non-woven fabrics go from fibre straight to fabric – this will be discussed in the next chapter.)

The way in which a yarn is produced is related to the texture, functional properties, thickness and weight of the final fabric. Yarn producers also look to trend and colour predictions when producing and developing yarns.

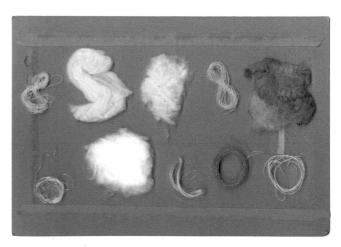

1

2

Yarn production

Denier
This is the thickness of man-made fibres. The higher the denier the thicker the fabric.

Cotton count
This is the numbering system for cotton yarns. The lower the number the thicker the thread. Sewing thread is normally 40.

Crêpe
This is highly twisted so that the yarn curls up, producing a crinkled surface in the finished fabric.

Bouclé
One yarn is wrapped around another yarn with a looser twist, which gives a pattern of loops, or curls, along its length. Fabric made from this yarn has a characteristically knobbly surface.

Slub
This is when some parts of the yarn are left untwisted.

Chenille yarn
Extra fibres are added into the twisted yarn to create this.

Nepp
Small pieces of coloured fibres are added that show up in the finished yarn.

During fibre production, synthetic fibres are put through a spinning process during which they are forced through small holes in a showerhead-style structure, creating long, continuous fibres called 'filament' fibres. Unlike natural fibres, manufacturers can control the thickness of the fibre during this process.

Staple fibres are short, natural fibres with the exception of silk, which naturally develops in a continuous length. Filament fibres can be cut to resemble staple fibres, so mimicking the properties of natural fibres. Synthetic fibres are cut down to become staple fibres when they are blended with natural fibres.

Spinning is also the name given to the process of twisting staple fibres together to make yarn. Yarn is twisted during the spinning process; the twist holds the short fibres together and contributes to strength. Yarn for weaving is tightly twisted to make it strong, while yarn for knitting is twisted more loosely to make it stretch. It also has better absorbency and a softer, warmer handle.

A single yarn is one yarn twisted, ply yarns are two or more single yarns twisted together. Two-ply yarn is two yarns twisted together and three-ply is three yarns twisted together. Ply yarns are stronger than single yarns.

Yarn can also be twisted and textured to enhance its performance or aesthetic qualities. Synthetic yarns can be heat set during manufacture to produce a texture.

Blending

It is common to blend yarns to provide optimum qualities in a fabric. Aesthetically a blended yarn may have a better handle and drape, and blending can also add function or reduce the cost of a fabric. Synthetic fibres are often blended with natural fibres to improve their qualities, for example, polyester mixed with cotton will produce a fabric with a natural handle that creases less. Lycra and spandex can be mixed with other fibres to give a stretch quality so that a fabric retains its shape with wear; this is especially suitable for performance sportswear. Blending can occur during fibre production, yarn formation or in the processes of knitting and weaving.

Dyeing

The colour of a fabric can inspire, motivate and attract a designer or consumer to a particular article of clothing. There are also other aspects that can enhance a garment, such as a particular novelty dyeing effect or specialty coating that creates a look or feel that is unique and desirable. A dye is a colouring matter that works as a stain. It is absorbed into the fibres of a textile; the colour is not as easily worn away as a colour applied to the surface such as pigment or paint. Colour can be applied with synthetic or natural dyes at fibre, yarn, fabric or finished garment stages of production. If a garment is to be dyed, first test it for shrinkage; dyeing often requires high temperatures to fix the colour properly and the heat can cause the fibres to shrink. It is also important to make sure all the parts of the garment will react to the dye, for example, the thread, zips, elastic and other trims. Fabric should be washed to remove any coatings before the dyeing process, as this will allow for better absorption. Remember that the original colour of the base cloth will affect the final dyed colour. Fabrics can be cross dyed for interesting effects, for example, a fabric composed of silk and viscose fibres could be dyed with acid and direct dyes respectively. Most dyes can be used for printing fabric when mixed with a thickening agent.

1

2

1   An example of a dye book.

2   A design from Alexander McQueen's S/S99 collection. Catwalking.com.

The water droplet test

Natural and synthetic fibres/fabrics contain impurities such as oils and starches, and may also have been subjected to finishes applied during production that can stop the successful absorption of dyes. A simple method for checking the purity of a fabric to be dyed involves applying a droplet of water to the surface. If it is quickly absorbed there are no or few impurities or coatings on the fabric. If the water remains on the surface the fabric will need to be washed before dyeing.

Natural dyes

Dyes originally came from soil, plants, insects and animals. For example, cochineal was obtained from the body of the female cochineal beetle and was used to produce red dye, while Tyrian blue was produced from shellfish. Some dyes sourced from natural dyes produce a subtle colour, but their light and colourfastness is not as good as synthetic dyes. Natural dyes use renewable sources, however, many need a huge amount of natural products to produce a small amount of dye. Also many natural dyes need mordants to fix the colour and these can be harmful to the environment. There are two kinds of natural dyes, adjective and substantive.

Adjective dyes need a mordant to help the cloth absorb the dye. The dye must form strong chemical bonds with the cloth to set the colour permanently. The mordant enters deeply into the fibre and when the dye is added the dye and the mordant combine to form a colour; since the mordant is thoroughly embedded, so is the colour.

Substantive dyes do not need mordants during the dyeing process to make the fabric light and wash fast. Indigo is the best known of these dyes.

Mordants
These prepare the fibre to receive the dyestuff and help the bonding. Most mordants come from minerals such as tin, chrome, alum (potassium alum), iron (ferrous sulphate) and tannin (tannic acid).

Natural mordants include mud, rushes, fungi, fruit peel and urine. The use of different mordants with the same dye can produce a variety of colours.

1–2 Viktor & Rolf S/S05. Half the collection was shown in shades of black and the second half in shades of pink. Catwalking.com.

1

2

Synthetic dyes

Towards the end of the 19th century, fabric manufacture expanded at a rapid pace due to the industrial revolution in Western Europe, predominantly in the UK. Great quantities of natural resources were needed to produce the dyes for the fabric. In some cases, the natural dyes were shipped from abroad, which was expensive and time consuming. As a result, chemists started to look at ways of producing synthetic 'copies' of natural dyes. At this time, a purple dye called Tyrian purple was used to colour cloth worn by royalty; it was a difficult and expensive colour to produce as it was extracted from the mucus of molluscs. Fortunately, a young chemist named William Perkin accidentally invented the first synthetic purple dye, which was called aniline

purple, or mauveine. His discovery made him very wealthy and paved the way for the research and development of other synthetic dyes. Today, synthetic dyes are developed continuously to improve their colourfastness and performance, and as a response to new fabrics that are being invented.

There are a wide variety of synthetic dyes formulated for different fabric types and for specific effects. Synthetic dyes tend to have a better light and wash fastness than natural dyes.

Basic dyes

Basic dyes were the first synthetic dyes to be developed, for example, Perkin's aniline dye. They have poor light and wash fastness and are not often used today except to dye acrylic fibres.

1

Acid dyes

The first synthetic dyes developed for wool were acid dyes. This class of dye has since grown into a large, diverse, versatile and widely used group. Some acid dyes may also be used for dyeing other protein fibres, including silk and also nylon or polyamide (at a higher temperature), which have a similar structure to protein fibres.

The term 'acid' refers to the fact that acid or an acid-producing compound is used in the dye bath. There are

different types of acid dyes including levelling and milling acid dyes. Levelling acid dyes are available in a range of bright colours and have good light fastness, but their wash fastness is only moderate. Milling acid dyes are also available in a range of bright colours and have good light and wash fastness, but are more difficult to apply correctly than levelling.

Direct or substantive dyes

These are suitable for dyeing cellulose fibres such as cotton and linen, but can also be used on silk, leather, wool and cellulose-mix fabrics. The first direct dye was called Congo red and was introduced in 1884. It was called a direct dye because it was the first dye to become available for colouring cellulose 'directly', without the use of a mordant. However, the addition of common salt (sodium chloride) or Glauber's salt (sodium sulphate)

improves the take up of dye. If wool or silk is dyed then acetic acid is added instead. Direct dyes are simple to use and come in a wide range of colours, albeit not very bright colours. The resulting dyed fabrics have poor wash fastness, so direct dyes are rarely used for printing.

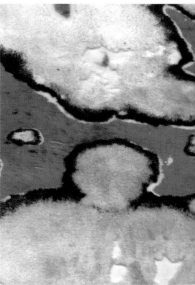

Dye classes

Different dyes are suitable for different fibres:

Cellulosic
Direct, reactive, vat, natural.

Wool
Acid, reactive (some).

Polyamide
Acid, reactive (some).

Acrylic
Basic, disperse (some).

Polyester
Disperse.

Cellulose acetate
Disperse.

Nylon
Acid, disperse.

Disperse dyes

Disperse dyes were introduced in the 1920s to dye acetate fibres, which were otherwise undyeable, with the notable exception of the natural dye logwood black, which was already being used on silk and wool. Nowadays disperse dyes are mainly used for polyester fibres, but are suitable for most synthetic fibres. They are applied at relatively high temperatures so are not suitable for use on fabrics that are mixed with wool as the wool may felt. Disperse dyes have brilliant light-fast properties.

Pigments

Pigments are used for printing fabrics when mixed with the appropriate binder or thickening agent. They are easy to apply and do not need to be washed afterwards to remove the binder. This, however, means the printed area can be slightly stiff to handle.

Reactive dyes

Developed in the 1950s, reactive dyes were the first dyes produced that chemically reacted with the fibre (usually cellulose) under alkaline conditions. The dye thereby becomes part of the fibre, rather than merely remaining an independent chemical entity within the fibre, which gives the fabric good wash and light fastness. Dyeing takes place in alkaline conditions normally through the addition of sodium carbonate. Common salt or Glauber's salt is also added, which helps the fabric take up the dye evenly. Variations in the amount of alkali and salt produces lighter or darker colours. Reactive dyes are suitable for dyeing cotton, linen and silk and are often used for printing. These dyes were first marketed by ICI in 1956 as Procion dyes.

Vat pigments

These are actually pigments that are insoluble in water. In order to apply them to fabric they must first be subjected to a process of chemical reduction known as 'vatting', which makes them soluble. They can then be absorbed into the cloth. The fabric is then exposed to the air or treated with an oxidising compound, whereby the reduced soluble form of the dye is reconverted to its original insoluble pigment form in the fibre and so is not liable to be removed by washing.

Multi-purpose dyes

Dyes produced for home use will dye most fabrics as they contain a mixture of dye types. They can be used in hot or cold water and some are ready to use in the washing machine.

1–2 Print samples by Mika Nash. These sample have been created by first dyeing the devoré velvet with direct dischargeable dyes. Discharge paste is frozen into ice cubes then placed on the fabric and allowed to melt. When dried the fabric is turned over and printed with devoré paste; the dye sets and also acts as a resist to the devoré.

Dyeing effects

Dyeing techniques can be used to create pattern. Certain techniques employ the use of resists that are applied to the fabric and act as a barrier to the dye. When the fabric is dyed the resist is then removed and the fabric is left with a negative pattern.

1–2   La Petite S***** A/W07. Silk satin and black organza dip-dyed dress. The design features a side seam that allows the organza layer underneath to be seen in between the sides of the baby-locked silk top layer.

3–4   Tie-dye textile samples by Furphy Simpson.

Tie-dye

Tie-dyeing involves tying twine around areas of the fabric before dyeing; the twine prevents dye from penetrating the cloth. When the fabric is untied and dried, undyed areas form a pattern on the fabric. Fabric can also be stitched before it is dyed. The stitches are pulled tight, which creates areas where the dye cannot penetrate. Fabrics can also be gathered or folded first, creating interesting effects. Tie-dyeing has an interesting history as it has been used since ancient times – the Japanese call it 'shibori' and the craftspeople of one region have developed numerous, beautiful designs.

3

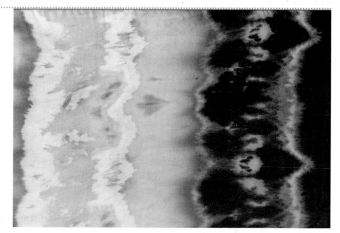

Starch and batik

Starch and wax can be used to paint and draw designs on to fabric. The starch or wax is left to dry and the fabrics are dyed. The starch is then flaked off or with the wax-resist technique (batik) the fabric is boiled and the wax melts off.

4

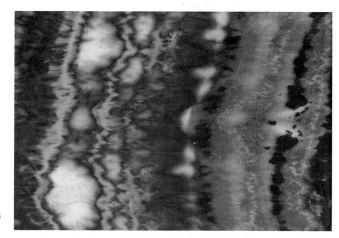

Ikat

The warp or weft threads are tied with twine and then dyed, leaving a pattern where the twine was. When the warp or weft is woven into cloth it creates an ombré effect. A double ikat is produced when warp and weft are both dyed and woven together.

Colour testing

Companies test textiles under specific conditions for their colourfastness. The colour in swimwear must be fast to seawater and the chlorinated water found in swimming pools. Equally a blouse worn next to the skin should not discolour as a result of perspiration. Because of the wide range of end uses for coloured textiles, many tests have been developed to assess fastness. Testing involves comparing a dyed sample that has been exposed to an agency, for example, to light or to washing, with an original, to assess accurately any change in shade or change in depth of colour. Changes are accepted up to an agreed level depending on the end use of the dyed material, but if these levels are exceeded the product fails the test. During washing fastness a coloured sample is tested with white fabric to assess the extent of staining. Fastness to light is, in fact, a measure of the ability of the dye molecule to absorb radiation without being destroyed. In a dye with poor light fastness the molecule will be broken down by the absorbed radiation. No dye is completely fast to light, but it should not fade appreciably during the life of the article that it colours.

Finishing processes

1   Junya Watanabe S/S08
     menswear jacket with a
     shrunken wash finish.
     Catwalking.com.

2   Emma Cook A/W07 knitted
     cashmere mix cardigan.
     Parts of the knit, including
     the collar, have been
     boiled to create an
     interesting effect.

Mechanical and chemical finishes can either take place at the fibre
stage of the development of the fabric or on the actual finished
surface of the textile. Processes can be used to add extra properties
to a fabric or garment for visual, tactile or functional effect. Finishes
can last the lifetime of a fabric or may wear off with time.

1

Basic finishes

These processes are necessary to prepare a fabric for dyeing or printing. Cleaning removes starch, dirt or grease after weaving or knitting. Desizing removes substances added to yarn before weaving to make the yarn stronger. Calico is not desized and as a result has a stiff handle. Scouring removes impurities from wool.

Bleaching cleans and whitens manufactured fabric and can improve the dyeing process. During the stentering process the fabric is pinned along its selvedge and stretched to realign the warp threads to their perpendicular position. You can see the pin marks down the selvedges of fabrics that have been stentered.

The milling process incorporates felting (a process where moisture, heat and pressure is applied to a fabric) causing wool fibres to matt to improve the handle of the cloth. The singeing process makes a fabric smoother as the fabric is passed over a flame and excess fibres are burnt off. Mercerization is usually used on cotton fabrics. Here chemicals are added to the fabric that increase the fibre's lustre. It also makes the fabric stronger and more susceptible to dye. Fabrics can be treated with optical brightening agents that are colourless fluorescent dyes; these react to UV light making white fabric look whiter.

Aesthetic finishes

Aesthetic finishes help give a fabric the right feel or look. A fabric could have a high-tech finish added to it that makes it look more modern or it may be that its finish could be a wash process that makes the fabric look older. Putting a finish on a mixed-fibre fabric can create interesting effects as the fibres may react in different ways to the finish. For example, one fibre may shrink on heat, while the rest of the fabric remains static therefore creating a crumpled or embossed surface. Chemical processes can change the tactile quality of the fabric, becoming soft and velvety or maybe papery and dry to the touch. Brushing the back of looped-back jersey sweatshirting will produce a fabric that now traps air and will insulate better. Felting occurs when a fabric is heated or mechanically manipulated, causing the fibres to shrink and distort the fabric. During the calendering process a fabric passes between heated rollers producing a flat glossy surface. A moiré finish also can be achieved using patterned rollers; the pressure and heat used produces a higher lustre.

12

STUDENT-FO

Kennefor

2

MADE IN ENGLAND
BY
Kennett & Lindse
ROMFORD ESSE

Pleating

1   Abercrombie & Fitch shorts
     with 'wash' effect.

2   Barbour waxed cotton coat.

Washing can be used to give fabrics a creased or crinkled effect. Fabrics can be randomly creased by washing and leaving them unironed. Creasing and fixing the fabric before washing can form crinkles in specific areas. Permanent crinkles and pleats can be achieved on most synthetics and wool fabrics through applying heat and shaping as the fibres are permanently changed. This can also work on fabrics that are a high blend of synthetic. The hand-pleating process involves the fabric being placed between two already pleated textured cards; the cards are then rolled up and put into a steamer. The resulting fabric takes the texture of the pleated cards. The Siroset process is used for wool and can be applied to specific areas of a garment to create a press line. A chemical is sprayed on to the front trouser leg and then steam pressed.

Issey Miyake signature pieces are made from thermo plastic polyester jersey. The garments are made first then pleated, which changes the garments' dimensions. As the garments are so stretchy due to the pleats there is no need for zips or buttons. The flat construction of the garments has reference to the kimono.

Washing

Stonewashing was a hugely popular finish in the 1980s and was the fashion style of choice for numerous pop bands of that era. Stonewashing is achieved with the aid of pumice stones, which fade the fabric, but it is difficult to control and can damage the fabric and the machinery used to finish it. Acid dyes were introduced to perform the same task and the effects are called snow or marble washes, but this type of process is not environmentally friendly.

Enzyme washes, or bio-stoning, is less harmful to the environment. Various effects can be achieved, depending on the mix and quantity of enzyme used within the wash. Enzyme washes can also be used to soften fabrics.

Garments can be sand- or glass-blasted using a laser gun to target specific areas where fading and distressing is required. Lasers can also be used to produce precisely faded areas on a garment.

1

Performance finishes

Chemical and/or mechanical processes can alter a fabric for a functional process. Fabrics can be flame proof, stain repellent, anti-static, non-iron, moth- or mould-proof and can even be treated to reduce UV ray penetration. The use of new microfibres is being developed in this field.

Bacteria

Chemical treatments can also control the growth of bacteria on a fabric therefore reducing odour. Teflon-coated fabrics also provide an invisible protective barrier against stains and dirt – useful for practical, easy-clean garments.

Breathable

Breathable waterproof fabric is produced by applying a membrane to the surface that contains pores big enough to enable perspiration to escape from the body, but small enough to stop moisture droplets penetrating. GORE-TEX® is a superior example of this kind of fabric. The GORE-TEX® brand was first developed as light, efficient insulation for wire on Neil Armstrong's early space mission. (Teflon is used to produce GORE-TEX®). It was then developed and registered as a breathable, waterproof and windproof fabric in 1976. It is now used widely for its properties in outerwear and sportswear.

Reflective

Laminates applied to a textile give the cloth a new property and function. They can be visible or non-visible, while holographic laminates reflect and refract the light.

Waterproofing

Fabrics can be waterproofed by applying a layer of rubber, polyvinyl chloride (PVC), polyurethane (PU) or wax over the surface. These fabrics are ideal for outdoor wear and footwear. Natural oils left in the wool of a fisherman's jumper acts as a naturally-occurring waterproofing layer.

Dyeing > **Finishing processes**

1 Prada womenswear A/W07 featuring experimental fabrics. The cardigan is made from plasticised mohair. The outer side of the knit is sprayed with a plastic substance then heat-pressed to give it a shiny and puckered appearance, although it remains soft and supple. The skirt's satin fabric is woven with elastic yarn. When heat-pressed, the elastic shrinks causing the fabric to crinkle and pucker. A herringbone fabric is fused with silk satin then through a needle-punching process areas of the herringbone are brought forward through the satin cloth.
Catwalking.com.

'It is a freedom to be able to make your own fabric while you are working.'

Sandra Backlund

In order to transform lengths of yarn into fabric to wear, the yarn must go through a process of construction; the two main fabric constructions are knit and weave. Other types of construction include crochet, lace making and macramé. Fabric can also be made directly from fibres and solutions. Tyvek is made from the matting together of fibres to make a paper-like fabric. Mass-produced shoulder pads are made from foam that comes from a solution. Leather and fur are probably the oldest types of 'fabric' that have been used by humans to clothe themselves. They are not constructed, but are cut from the animal as a skin.

When looking at fabric construction it is important to consider what properties a certain technique will give to a fabric and ultimately the finished garment. A knitted fabric will tend to be used for its comfort stretch and ease of fit. A woven fabric may be used when a garment needs structure and stability, whereas a lacy fabric could be used for its decorative qualities. However, a knitted fabric can be structural if knitted and then felted. A woven fabric can be stretchy and comfortable if woven with Lycra and can also be decorative if produced on a jacquard loom.

1

Weave

Structure variations

There are many variations in the structure of plain-weave fabrics:

Ribbed
This is created by grouping warp or weft yarns or using thicker yarns in areas of the cloth.

Basket weave
Is a loosely woven fabric achieved by alternately passing a weft under and over a group of warps so the weft lies over the warps. This is repeated to produce a square pattern in the fabric.

Seersucker
Here the warp is held at different tensions, one set of warp yarns is held tight with the remaining yarns held slack. As the weft is inserted across the warp a puckered effect is created by the looser warp yarns.

A woven fabric is made from a warp that runs down the length of a fabric and a weft that weaves across the breadth of the fabric. The warp and weft are also known as the 'grain'. If the grain is not at 90° the fabric is said to be 'off grain' and may not hang or drape properly, which can cause problems when making up a garment. The warp is stretched on to a loom before weaving; this means there is more 'give' across the width of the fabric where the weft is woven across. The warp is sometimes coated with starches to increase the strength of the yarn; these starches are washed out in a finishing process when the fabric is woven. The loom traditionally had a shuttle carrying yarn back and forth under and over the warp yarns. This process can still be seen in today's production methods. Newer shuttleless looms use air or water jets to propel the weft yarn across the warp at incredibly fast speeds. These machines are a lot quieter than traditional looms. The weft yarn is not continuous, but is cut to length before it is passed across the warps. Looms can be circular, producing a tube of fabric, and also double-width, producing two widths of fabric at the same time – denim is often woven in this way.

The way the warp and weft are woven together produces a variety of fabrics. The three main types of weave construction are plain, twill and satin.

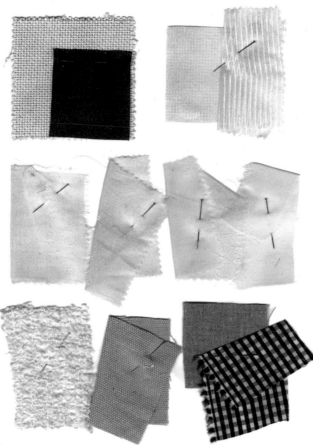

2

Plain weave fabrics

Calico
A plain cotton fabric that has not gone through a finishing process. It therefore still contains starch from weaving and has a slightly stiff handle.

Canvas
A heavyweight tightly-woven cotton fabric.

Chambray
A medium-weight fabric with white warp yarns and coloured weft yarns of cotton or cotton mix origin.

Chiffon
A lightweight soft fabric.

Gingham
Features a small check weave structure usually made up of two colours.

Muslin
A lightweight soft-handled plain weave, usually white or unbleached cotton.

Organdy
A sheer, crisp lightweight cotton.

Organza
A sheer, crisp, lightweight cotton made from filament yarns, for example, silk or synthetic yarns.

Voile
A lightweight fabric made with two-ply warp in cotton, cotton mixes or man-made fibres.

Plain weave

1   Marios Schwab A/W07 layered silk-chiffon dress. The chiffon is a plain weave construction.

2   Plain weave structures. Shown top row (left to right): plain weave, plain weave with chintz finish; basket weave waffle, rib.

Middle row (left to right): cotton voile, georgette; silk organza, chiffon.

Bottom row, (left to right): searsucker canvas; chambray, gingham.

Plain weave is constructed from a warp and weft that is similar in size. During the weaving process the weft is passed over alternate warp threads to create the fabric and is usually closely woven. Basic plain weaves have a flat characteristic and are good for printing and techniques like pleating and smocking. Using different yarn weights and tensions creates variations to the plain weave.

Twill weave

1   Peter Jensen A/W07 wool tweed jacket and skirt.

2   Twill and satin weaves. Shown top row (left to right): silk cotton twill, twill, cotton gaberdeen, drill.

Middle row (left to right): denim, tweed, herringbone, houndstooth,

Bottom row (left to right): polyester satin, silk satin, cotton sateen.

During the twill weave process the weft is woven over at least two warp threads before it goes under one or more warp threads. Where this is staggered down the length of the fabric it produces diagonal lines on the surface of the fabric; these lines are called wales. The wale can run at various degrees across the fabric; a regular twill runs at a 45° angle, while a steep twill runs at more than 45°. Twill lines or wales can also run from left to right, a right-hand twill, or right to left, a left-hand twill. Wales can show on one side of the fabric or can show equally on the front and back of the fabric. Twills are usually closely woven and are strong and hardwearing fabrics.

Twill weave fabrics

Chino
This has a steep twill and is made from combed or two-ply yarns.

Denim
Usually made from yarn-dyed cotton or cotton blends.

Drill
A dyed medium- or heavyweight fabric.

Herringbone
An even-side twill where the wale regularly reverses to form a chevron pattern.

Tweed and houndstooth
Twills using different coloured yarns and weave structures to create pattern.

1

Satin weave

Satin weave has visible sheen and feels smooth, due to a tightly woven weave structure that allows yarn to lay across the surface of the fabric. The warp is woven to lie on top of the weft or vice versa. Satin weave fabrics are often used for lining as they glide easily over other garments.

3

3   Peter Jensen A/W07 silk satin top incorporating pearl necklace detail.

4

Satin weave fabrics

Double-faced satin
The front and back have a smooth satin line finish as the fabric is woven with two warps and one weft.

Crêpe-back satin
The fabric is woven with a high-twist crêpe yarn that shows on the back and a lower twist yarn that shows a smooth satin finish on the front.

Sateen fabrics
These are made from spun yarns, usually cotton.

Satin fabrics
These are often made from filament yarns with low twist.

4   Spijkers en Spijkers S/S06 silk satin panelled dress.

Weave > Knit

Other weave structures

The three basic weave structures can be varied to produce more complicated weave structures.

Pile fabrics

These are woven with extra yarns to the basic warp and weft, which create floats or loops that can then be cut, for example, corduroy, or left as loops, as in towelling.

Double cloth

This is the result of weaving two interconnected cloths at the same time. Velvet is commonly woven as a double cloth – that is, cut apart after weaving to produce two fabrics that are the same. Double-cloth construction can also produce a fabric made of two quite different qualities, patterns or colours. This kind of fabric is reversible so that either side can be used as the outer layer of a garment. Sometimes the fabric is connected all over or it may be just connected in some parts creating pockets and puckers.

Pile fabrics that show pattern

Corduroy
Extra weft yarn is woven leaving regular floats, which are cut and brushed up to form piled lines working down the fabric that can be of varying thicknesses. Fine lines are known as needle cord, large lines as jumbo cord.

Towelling
This is created by slack tension weaving, which creates loops on the surface of the cloth in a similar way to the construction of seersucker as discussed earlier.

Velveteen
A fabric with an all-over cut pile, usually made from spun yarns.

Other weave structures

Brocade
Usually using different coloured yarns to produce a highly patterned fabric with many floats on a plain, satin or twill base.

Damask
Usually constructed in one or two colours in a similar way to brocade.

Jacquard
Creates patterns and textures through a complicated weave system in which warp and weft threads are lifted or left. Often long floats occur on the fabric that can catch and snag.

Dobby
These weaves have small repeated geometric patterns in their structure.

Pique
The dobby weave is used to create a surface pattern.

Spot weaving
Yarns are added to the basic warp and weft to create pattern and texture in places.

Waffle
A dobby weave that produces a honeycomb pattern.

Engineered weave

It is interesting to consider engineering pattern, colour or function in a weave. An elastic yarn could be woven across the width of a fabric to change its construction. The elastic when released would constrict, bringing the fabric in. This could then be cut and incorporated into a garment that when placed on the body would tighten to fit it in specific areas. Computer-aided weave can produce fabrics with many layers and surfaces. In jean production sometimes the denim is woven narrow so that the selvedge edge can be incorporated into the trouser leg. It is the finished edge to the inside seam.

Shape-memory alloys, such as nickel and titanium, can be woven into fabrics so that the fabric regains its shape with the application of a certain temperature. This principle is used in the production of bras; the wire in the cup retains its shape and bra structure when the bra is put through the washing machine.

2

Double-cloth fabrics

Melton
Usually wool produced in the double-cloth process, but not cut through afterwards, which produces a heavy fabric suitable for outerwear.

Velour
A woven fabric or felt resembling velvet. Produced using the same process as velvet construction, but made usually from cotton yarn.

3

1   Other weave structures. Shown left (top to bottom): silk, velvet, cotton velvet, jumbo cord, needle cord.

Middle (top to bottom): ripstop dobby patterns, cotton spot weave with cut threads, wool double cloth.

Right row (top to bottom): jaquard, moleskin, resin-coated cotton, crêpe.

2   Eley Kishimoto A/W07 dress featuring an engineered weave in the design.

3   Evisu jeans with a selvedge edge used in leg seam.

Knit

Knitting dates back to the Egyptians, but was really developed into an industry in Europe in the early 16th century. Knitting machines were developed in the second half of the 18th century due to the demand for patterned stockings and following the invention of the rotary knitting machine, tubular knitting could be produced for hosiery.

In the 19th century British sailors and fishermen developed styles of knitting that incorporated pattern and texture that are still well known today. Fishermen's Guernseys or 'ganseys' originated in the Channel Islands in Guernsey and at one point a fisherman's region of origin could be identified by the pattern of his Guernsey. Texture was important, often incorporating cables that looked rather like the fishermen's ropes. An Aran knit was originally cream and heavily embossed with cable, honeycomb, diamond and lattice patterns. The patterns were handed down from generation to generation often just visually rather than being written down as a pattern.

Fair Isle is a term that is used to describe patterned knitting in multicolours. It should, however, really be used to describe the specialised colours and patterns used in Shetland knitting; Fair Isle is an island south of Shetland. Shetland is positioned between Scotland and Norway and the influence of a folk-style motif is apparent in Fair Isle patterns; the motifs tend to work in bands on the knit and are not random. Shetland sheep produce fine soft-quality wool that is not clipped from the sheep, but plucked by hand. The wool is available in a variety of colours such as white, cream, fawn, grey and black. The wool is also dyed with lichens to achieve soft rose, pale yellow and purplish browns. Patterns from other European countries tend to be more bold and graphic than British designs. The Bjarbo pattern from Sweden was traditionally worked in red and blue on cream ground. Scandinavian designs also use small figures on a light ground and an eight-pointed star was common in Norwegian designs.

In the late 19th and early 20th century fashions changed and women and men began wearing jumpers and cardigans and other knitted garments for day, evening and more importantly for sports. With the increase in leisure time in the 20th century, knitted fabric has become increasingly important for sportswear as it is stretchy, comfortable and absorbent.

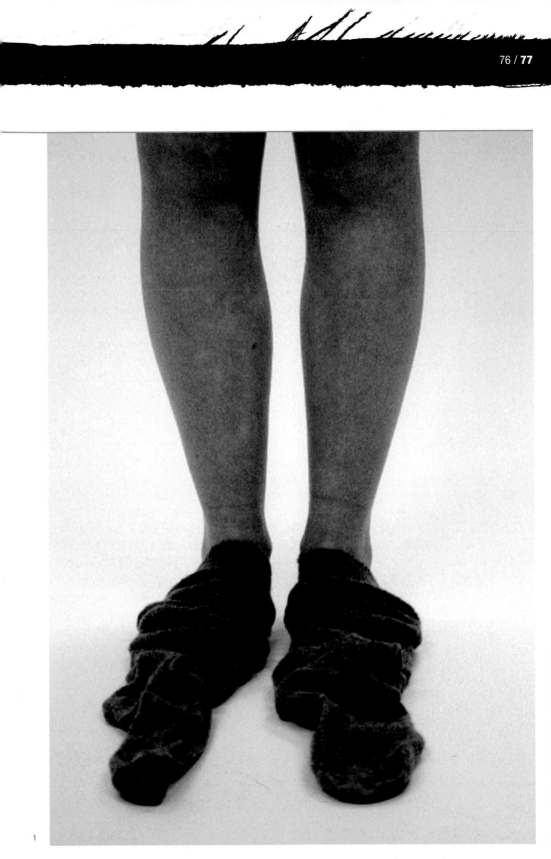

1

Knitted fabrics

Knitted fabrics are constructed from interconnecting loops of lengths of yarn, which can be knitted along the warp or weft giving the fabric its stretchy quality. Horizontal rows of knit are known as 'courses' and vertical rows are known as 'wales'. Weft knitting is created from one yarn that loops and links along the course; if a stitch is dropped the knit is likely to ladder and run down the length of the wale. Hand knitting is a prime example. Warp knitting is more like weaving, with the construction being more complicated and the fabric less easy to unravel.

Knitted fabric tends to be comfortable to wear as it is stretchy, but this can also mean that it can stretch out of shape and can shrink with heat, especially if it is made from wool. Knitted fabrics tend to be more prone to pilling than woven fabrics. This is because loosely spun yarns are often used for knitted construction and

these pill more than tightly spun yarns. Different thicknesses of knitting can be produced according to the stitch used, the size of the needles and the thickness or the count of the yarn. Knit finishes can change the quality of the fabric. For example, softeners can be used to improve the handle of the cloth. Washing at a high temperature and using friction can cause woollen knit to felt, which in turn causes the fibres to matt together making the knit denser and less stretchy.

Gauge
The number of rows and/or stitches per length/width of a knitted fabric. For example, five stitches per inch.

1

Basic stitches

Texture and pattern can be created by knitting with different needles, yarns, colours or stitches. Stitches add decorative quality, but they can also benefit the physical quality of a knit, for example, the use of tuck stitch or cable increases the density of the fabric. It is important how the swatch looks, but also how it feels and how it drapes. Experiment with stitches, techniques and tensions to create something unique. Knit can be embroidered to highlight areas and beads can be inserted into stitches for a more decorative effect.

Slip or float stitch

These are used to create patterns or change colour within a piece of knitting. The yarn is stitched then floats across the back of the fabric before it is used again. It is best not to allow the floats to be too long as they tend to catch and cause snagging. These floats also restrict the stretch of the fabric and make the finished knit heavier. Fair Isle and jacquard incorporate this method to create pattern. The wrong side of the fabric can be used as the right side and the floats can be cut, again for decorative effect.

2

Intarsia

Pattern is created with yarn changes, but the yarn does not travel across the back of the fabric creating floats, which means that larger colour blocks can be used in a design.

1   Sonya Rykiel S/S08. Knitted garment featuring graphic details of a macintosh coat, including a belt, collar and pocket.

2   Winni Lok knitted sweater using the right side of the fabric on the front of the jumper and the wrong side with floats on the back.

3   Vintage Sonya Rykiel Marino wool intarsia pattern jumper.

3

Weave > Knit > Other forms of construction

Tuck stitch

The stitch is held on the needle as the rest of the piece continues to be knitted creating a pulled or tucked effect. Small honeycomb patterns to larger bubbled or puckered effects can be created and both sides of the fabric can be used showing either a bubbled look or indent.

Partial knitting

Needles are selected so some stitches are knitted and others are held on the needles. This is used for decorative effects and also for shaping garments for a dart or flare.

1   Vintage Tao Comme des Garçons 100% polyester blister effect top.

2   Kenzo cardigan.

3   Marc by Marc Jacobs jumper.

4   Clare Tough hand-knit.

1

Lace stitch or stitch transfer

A stitch is transferred to another needle, creating a hole as the knitting continues. Single or groups of stitches can be transferred from needles by hand using a transfer tool to create interesting patterns. Lace knitting has been produced in Europe since the 15th century, but did not become popular until the 18th century when fine cotton was imported from the East. It was known as white knitting as the yarn used was predominantly undyed.

Cable knit

Stitches are transferred across a
knit creating raised twisted groups
of stitches. Cables are used more
commonly in hand knitting as raised
three-dimensional designs can be
created, however, they can also be
created on the V-bed machine.
Wool yarn is good for this technique
as it is elastic and will allow stitches
to stretch during transfer.

3

Inlay

A yarn is woven or laid on to the
knit and the stitches catch it down.
The laid yarn does not make loops
as normal knitting does, so it tends
to make the knitted fabric less
stretchy. Yarns that would normally
be too thick, thin or maybe too
textured to knit with, can be
successfully incorporated into
knit using this method.

Hand knitting

Hand knitting can produce a variety
of weights of fabric and has its own
'home-made' character; it is
especially suited to very heavy knits
and cables. It is possible to create a
fabric very quickly with thick yarns
and large needles. It is also a very
transportable means of constructing
a fabric as it can be carried around
and worked on in any location.

4

1–2   Dubied and domestic
      knitting machines.

3   Shown from top to bottom:
      airtex, single jersey stripe,
      interlock, piqué, loop back
      sweatshirting, fleeceback
      sweatshirting, rib.

4   Christian Wijnants 2007,
      silk stretch jersey dress
      with marble wall digital
      print.

Machine knitting

Originally, knitting was produced by hand, but this developed into machine knitting for mass production. Yarn can be knitted flat or circular and as a length of fabric or fashioned to fit – knitted socks are an example of fully fashioned machine knitting.

Punch cards are used on domestic machines to select needles more quickly than by hand in order to create pattern and texture. The card is fed into the machine and row-by-row selects needle positions. Pre-punched cards can be bought or cards can be punched for specific designs. Computer-linked jacquard and knitting machines can produce very intricately patterned fabrics from drawings and photographs through complicated needle selection, and designs can also be worked across a large area. The use of CAD also allows knitting designs to be changed quickly and respond to fashion trends.

It is important not to use a yarn that is too thick for the capacity of the machine, otherwise it will jam up or the yarn will snap. A thicker yarn could instead be laid into a design. Certain yarns that are darker may knit in a different way to paler yarns as the dye has altered the properties, making them less elastic. Wool is one of the best yarns to knit with as it recovers its shape well after stretching or distorting. In contrast, cotton can be hard to knit with due to its inelastic quality and it is more liable to break.

Shaping aids are used with domestic machines to allow the knitter to quickly and simply create shaped knitting without having to manually count stitches and rows.

Machines have beds of horizontal needles with each needle producing a column of stitches (wale) in the fabric. Single-bed machines have one set of needles all working in the same direction producing stocking stitch. Double-bed or V-bed machines have two sets of needles set opposite each other and produce double-knit or rib fabric. Examples of electronic machines are Stoll, Shima Seiki and Protti. Dubied is a hand-operated knitting machine.

One-needle bed

Single-jersey knit has a front 'knit' and a back 'purl', and is produced when using one bed of needles. This fabric can be heavy- or lightweight, can ladder and run, and tends to curl when cut. Sweatshirting is a heavier knit, the back of which is looped and brushed to achieve a fleeced effect.

Two-needle bed

Interlock knit or double jersey is produced with a double row of needles and the knit looks the same front and back, both showing a knit stitch, or a 1x1 rib.

Ribs and other textured knitting are produced using two beds of needles knitting alternate knit and purl stitches. Ribs can be used to finish garments on the cuffs or waistband where a garment needs to be gathered in. Due to their construction they have greater stretch. Ribs can

also be used to produce a whole garment. A 2x2 rib has two columns of knit stitches and then two columns of purl stitches alternating across the fabric, 3x3 would have the corresponding number of stitches and so on. Rib knits are more elastic than single-jersey knits.

3   4

Circular knitting

Circular machines produce a tube of knitting. They are very fast as they continuously knit and one row can be started before the last row has finished. The fabric, however, can twist due to the manufacturing process. If a flat piece of fabric is needed, the tube is then cut through and the edges sealed before being processed as it has no selvedge and would otherwise run.

Warp knits

Warp knits are usually created on a flat knitting machine and this is the fastest way to produce fabric from yarn. The main makes of machine are Tricot or Raschel. Tricot knits tend to use fine yarns and produce smooth, simple fabrics, whereas Raschel knits are more textured and have open work designs in heavier yarns. Fine nets, laces and powernet fabrics are produced on a Raschel machine.

New research in warp knitting has resulted in the creation of seamless garments. Issey Miyake is known for his A-POC (a piece of cloth) tubular clothing where garment shapes are cut out of a knitted tube length, each garment featuring cut lines that when cut do not run or ladder. The wearer interacts with the garment and can customise it using the cut lines to their specific requirement.

Weave > **Knit** > Other forms of construction

Other forms of construction

Types of lace

Alençon
A needlepoint lace. A fine corded pattern worked on a mesh background.

Chantilly
A very delicate intricate lace often featuring flowers or vines.

Cluny
French bobbin lace.

Rose Point
A needlepoint lace with elaborate raised patterns.

There are many other fabric constructions other than knit and weave, and they can be used to produce a variety of fabrics ranging from decorative, handcrafted looks to functional and technologically advanced creations.

Knotted and twisted

1   A vintage crochet top.

2   Different lace samples.

These techniques can be seen as more craft based. They involve a yarn or yarns that are twisted and knotted together to produce fabrics that are decorative and have an open structure.

Crochet

The word crochet comes from the French word *croc*, meaning hook. Stitches are made using a single hook to pull one or more loops through previous loops of a chain. This construction can be built up to form a patterned fabric. Different from knitting, crochet is composed entirely of loops made secure only when the free end of the strand is pulled through the final loop. Crochet hooks vary in size so they can be used to produce different structures of fabric. Fine needles and yarn create a lacy fabric, while thicker needles and hooks create a more solid fabric. Textures can be created by either wrapping yarn around the hook before it is stitched or by working stitches on top of existing stitches. More open-work fabric with holes and spaces can be created by forming bars and lines of stitches. Fabrics can also be made by building up rows of stitches or from working in the round, from a circle out.

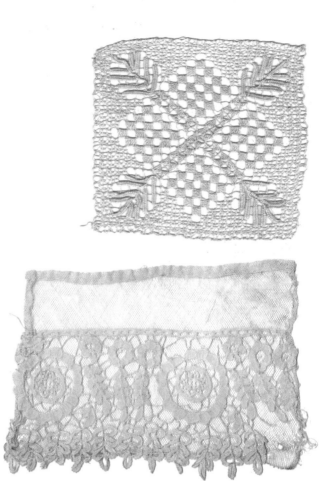

2

Macramé

Macramé is constructed through the ornamental knotting of yarn, giving the fabric a 'handcrafted' appearance.

Lace

Lace making produces a fabric that is light and open in structure. The negative holes in lace are as important as the positive stitches in the overall pattern of the fabric. Needlepoint lace is based on embroidery techniques and bobbin lace is based on braiding techniques. They were both developed in the late 16th century and were the most expensive type of ornamentation at the time. In Europe lace production mainly came from Italy, France and Belgium and the lace was named after the region it was produced in. Irish crochet lace originated from the Italian needlepoint lace of the 17th century and from the 19th

century, Ireland became the main producer. Lace became very popular again in the early 20th century, with patterns often depicting organic shapes and insects in the art nouveau style.

Originally lace was made by hand, but now it is mainly produced using the Levers machine. Lacy knits can be created on Raschel knitting machines and lacy embroideries can be produced using the Schiffli embroidery machine.

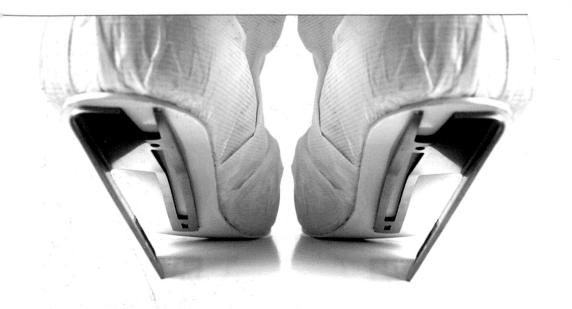

1

Non-woven constructions

Non-woven fabrics can be used for
fashion garments, but are also used
for linings, padding and the interiors
of shoes and bags. Due to their
construction, non-woven fabrics
have no grain and do not fray or
unravel in the same way as woven
fabrics, which makes them
eminently more suitable for
garments or accessories that need
to be more hard-working and
reliable. Other non-wovens are
being developed for future fabrics.

Chemical and mechanical

Compressing fibres together with
the use of heat, friction or chemicals
can produce fabrics. Examples of
this are felt, rubber sheeting and
Tyvek. Tyvek, produced by DuPont,
is made by matting fibres together –
almost like the way paper is made.
It also has a coating that makes it
tear-proof, water-resistant,
recyclable and machine-washable.

Fabrics can also be made from
solutions; foams and films are
examples of this.

Sprayed fabric

Manel Torres has developed a
way of producing a textile by literally
spraying the fibres from a can; in
this way a fabric can be sprayed
directly on to the body. Areas can
be built up to be thicker than others
and there is no need for fastenings
as the garment is cut off at the end
of its use. The fabric adheres to the
form of the body or if a shape is
needed then moulds can be placed
underneath to shape the garment.
In this way, Torres is combining
design with chemistry.

3

3D forming

1–2   Tyvek, leather and stainless
steel slip-on boots by
Marloes ten Bhömer.
These boots look into the
aesthetics of destruction,
in order to create a new
silhouette form. The front
of the boot is distorted and
dented. Normally a slip-on
boot is very straight at the
back so your foot can slide
into the boot. The way the
patterns have been cut
enables the back of the
boot to hug the calf
and leg.

3   Manel Torres sprayed
fabric dress.

Using computer technology, objects
can be created three-dimensionally.
So far this technology of three-
dimensional printing has grown out
of rapid prototyping, which has
been used with great success in
the engineering industries. Sports
footwear design has benefited from
this computer technology. This
technology allows a three-dimensional
map of the body to be taken
digitally and a garment produced to
perfectly fit the shape of the body.

Knit > **Other forms of construction**

1   Emma Cook A/W07. Birds were the inspiration for this collection. From feather imagery, which is printed in pigment and flock, to a dress that is completely embellished in feathers; producing graduated colour and pattern.

Once a fabric has been constructed, it can be enhanced or altered with the application of different kinds of surface treatments. Pattern, colour and texture can be added to fabric through such treatments. Techniques include print, stitch, fabric manipulation, beading and embellishment. It is important to consider the type of technique that best suits the fabric you are working with. A loose-weave fabric would be suitable for drawn work (discussed later), whereas a tightly woven fabric with no pile is the easiest to print on. It is wise to consider the function of the fabric, for example, it could be made light reflective through the application of a print finish or quilting could be incorporated to improve heat retention. Some techniques might be interesting as a sample, but might not be applicable to a fashion garment. Consider if you are making the fabric unstable by deconstructing its structure perhaps through devoré or drawn work, or if the resulting fabric becomes impractical and heavy as a result of too much embellishment. Consider whether the fabric can be successfully worn again, whether it can be washed or dry-cleaned. It is a good idea to test a sample of the fabric you are creating by finding a dry-cleaner who is willing to trial small samples for you.

Print

1   Peter Pilotto A/W07
     collection.

2   Examples of hand-painted
     fabric and inspiration.

3   Printing blocks.

Print can be applied to a fabric through the techniques of screen-, block, roller, mono, hand or digital printing. Pattern, colour and texture can be achieved by printing with a variety of media, including pigment, dye, flock or glitter.

Processes

Although certain processes are more applicable to specific fabrics, it is important to experiment as a fabric might react in an unexpected way. Always make notes of the processes you are trying out to refer back to later.

Consider how you work with pattern on the fabric; the scale, proportion of colour, placement and repeat will all affect the overall look of the fabric and ultimately the garment it is to be used for.

1

2

Hand painting

As the name implies, mono printing produces a single, unique print. Inks are applied to a surface that is then transferred to the fabric, in reverse, to make a print. Hand painting is made directly on to the fabric using one of a number of tools, such as brushes and sponges. Hand painting gives a 'hand-made' feel to a piece of fabric, but can be a slow process for producing a long length of fabric.

Block printing

Block printing is one of the earliest forms of printing. A design is applied to a hard material – for example, wood, lino or rubber – via embossing or by cutting into the surface to make a negative image. This block can then be coated with ink and applied to the fabric with pressure to form an imprint. In 1834 Louis-Jerome Perrot invented the mechanisation of woodblock printing allowing multicoloured designs to be printed. The Perrotine printing process enabled the mass production of printed cloth.

Print > Embroidery and fabric manipulation

Roller printing

The flat copperplate-printing process was introduced in the 1770s making the printing of large repeats with fine engraved details possible. They were mainly one colour with extra colours introduced through hand-block printing or hand colouring. This was developed into the roller-printing machine and was patented by Bell in 1783. This meant fabric could be printed in a continuous length and mass produced. Printing multicolours through this process was developed shortly after.

1–2   Marc by Marc Jacobs A/W06 dress. The fabric has been printed using a rotary method, each colour, including the dark background, is printed. A rotary method ensures no print joins can be seen.

3   A digitally-printed dress design by Cathy Pill.

2

Screen-printing

Screen-printing requires a design, ink, squeegee and a 'silkscreen' – that is, a piece of silk stretched evenly across a frame. The first step is to make a stencil of the design, which is applied to the screen, blocking the silk so the ink can only pass through the 'positive' areas of the design. The screen is placed on the fabric and the ink is pulled through the screen evenly with the squeegee, leaving a printed image on the fabric. The print is then fixed on to the fabric with heat so that it will not wash off. Multicoloured designs are created through the use of different screens for different colours. The silk-screen process was used as far back as the 17th century. Nowadays the silk screen is made from a tougher nylon or polyester mesh and the stencil is processed using photographic emulsion.

Rotary

Rotary screen-printing involves a series of revolving screens made of fine metal mesh, each with a squeegee inside that forces the print paste through the mesh on to the fabric. The design is either laser etched from a computer on to the metal mesh or exposed in a photographic process. The rotary process is much quicker and more efficient than flat screen-printing.

Transfer printing

A design is printed with disperse dyes on to a transfer paper that is then left to dry. The paper is subsequently placed on the fabric, dye side down, and heat and pressure are applied so that the design is transferred on to the fabric. The transfer paper cannot be reused as all the dye has been transferred. The sublimation process of transfer printing ensures the dye penetrates the fabric rather than sitting on top of it. This gives the cloth a good handle and also does not affect the fabric's ability to breathe. This method is used on synthetic fabrics; however, non-synthetic fabrics can also be used, but they must first be prepared with a coating.

1

Digital printing

Inkjet printing for textiles is very different from the other types of printing already discussed because of the non-contact mechanics of the print head, but also the means by which the individual colours of a design are produced. Ink is directed through nozzles as a controlled series of drops on to the surface of a fabric, printing line by line. Usually a set of inks is used consisting of at least three or four primary colours, namely cyan (turquoise), magenta, yellow and optionally black, the so-called CMYK inks. As most inkjet printers were originally designed for paper printing, technical specifications are more related to those used in the reprographics industry than to those that a textile printer would normally employ. Reference is usually made to inks rather than dye solutions, pigment dispersions or print pastes. Similarly print resolution is usually defined as dots per inch (dpi) or lines per inch (lpi). The two main digital fabric printers used today are the Stork inkjet printer from the Netherlands and the Mimaki textile digital printer from Japan.

Technological improvements are enabling manufacturers to use pigments rather than dyes when making inks for textile printing using digital technology. Pigments are intrinsically more light fast and wash fast than dyes, and are often less expensive. Unfortunately, pigmented inks tend to flow less well than dye-based inks. This is important when delivering ink through a nozzle. Furthermore, technological advances in digital printing have led to improvements in the way that pigment-based inks adhere to the surface of the fabric. Pigment-based inks can be printed on to a broad variety of fibres and fabrics, whereas dye-based inks are restricted to specific types of fibres and fabrics.

When using dye-based inks the fabric must first be prepared with a chemical coating. The fabric is then printed and goes through a fixation process using steam to ensure that the ink adheres to the fabric. The fabric is then washed to remove the coating. Pigment-based inks do not need to go through this fixation process, which makes pigment-based printing more economical in terms of running costs. Unlike dye-based inks, pigmented inks do not require a solvent to dissolve the colourant. Such solvents are often based on volatile organic compounds, which means that dye-based inks tend to be less environmentally friendly than pigmented inks.

Digital printing allows the textile designer to work straight from the computer to cloth with no need for paper designing. Very high-definition imaging can be achieved and many colours can be printed without the need for numerous screens. Infinite layers can be built up within a design and clever three-dimensional imagery can be achieved. The repeat can be any size for a design and is not restricted by screen size.

3

Printing media

It is important to choose the correct media for successful textile printing. The media must fix to the cloth correctly and have a good handle for fashion purposes. Colour and pattern are achieved through the application of inks and pigments, and texture can also be added to fabrics through certain printing methods. Chemicals can be used to produce a 'relief' effect on the surface of the fabric or to 'eat away' at it to create a deconstructed look.

1

Inks

To print a colour a dye is used with an oil- or water-based thickening agent, which stops the dye from bleeding in the design. An oil-based ink is more opaque and heavy and tends to sit on the surface of the fabric. These inks are available in a range of colours and finishes, including pearlescent, metallic and fluorescent. Water-based inks leave fabrics with a better handle as the thickening agent can be washed out after the fabric has been printed and fixed. When printing on a stretch fabric a stretch auxiliary is sometimes added to the ink to improve the print quality so that it does not crack when stretched.

Types of printing media

Luminescent
Invisible in daylight, visible in infrared or UV light.

Fluorescent
Colours that glow in daylight or UV light.

Glitter
Fine monochrome and polychrome glitter inks.

Holographic
Interference high-contrast, viewing-angle dependent.

Hydrochromic
Textiles that respond to water.

Opaque
Colours that are not transparent. These are good for printing on dark fabrics.

Pearlescent
Soft, viewing-angle dependent polychromatic colour effects.

Phosphorescent
After charging with light, this glows in the dark.

Piezochromic
Textiles that respond to pressure.

Polychromatic
Colour-changing effects.

Reflective
Direct reflection of visible light.

Thermochromic
Inks that change colour with temperature.

Transparent
Colours with low opacity.

Discharge printing

A fabric can also undergo 'discharge' printing. First, the fabric must be dyed with a dischargeable colour. (You can see if a dye is dischargeable by looking at the manufacturer's dye information.) The fabric is then printed with a substance that bleaches away or 'discharges' the dye. Discharge printing is useful if a pale-coloured image is required against a dark background.

Devoré paste

Fabrics constructed with both natural and synthetic fibres within the warp and the weft can be printed with a devoré paste. When heated, the paste burns away one of the fibres, leaving behind a pattern where the other fibre remains.

Flock, glitter and foil

Fabrics can also be printed with glue then heat-pressed with flock paper. The flock adheres to the glue, creating a raised 'felt' effect. Glitter and foil can be similarly applied to produce special effects.

2

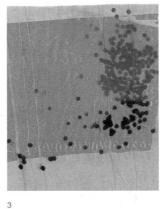

3

4

Puff

When printed and heated the ink expands on the surface of the fabric. Expantex is a brand of chemical that produces an embossed effect on fabric and has a rather rubbery texture. Three-dimensional qualities can be achieved by printing puff on the back of fabrics that are light or drape well. The puff distorts the fabric creating three-dimensional effects on the right side of the fabric.

5

1 Jean-Pierre Braganza A/W07 wool felt coat with plastisol print.

2 Clear discharge paste has rendered parts of this denim sample pale blue.

3 An example of the devoré print technique.

4 Digitally printed cotton with hand-applied bronze foil in parts.

5 Tata-Naka A/W07 grey marl jersey dress with nylon net. The top part of the dress features puff print on silk organza.

Print > Embroidery and fabric manipulation

Pattern

Textile designs have over the years been categorised into styles. It is important to have an understanding of these styles so that you can communicate with other designers or clients. The styles also show the wide variety of prints available.

Floral

Flower motifs. They are the most popular style and are reinvented each season. Leaves and grasses also go into this category, but fruits, vegetables and trees are classified as conversationals. Bouquets are tight ties of flowers whereas a spray is a looser, more free-flowing tie of flowers. Sprigs are small single stems of flowers.

Chintz

This fabric is typically glazed in appearance. Originally the fabric would have been rubbed with wax, starch or resin as it was thought that this would help repel dust and dirt. More recently the glaze is achieved with mechanical calendars. The floral pattern is the print style most associated with chintz.

Toile de Jouy

During the 18th century the town of Jouy in France produced printed cotton fabrics, especially a fabric depicting the printing process of the factory set in an outdoor landscape that was put into repeat. The name is now associated with a design that represents a repeated landscape or pastoral scene. The design is usually printed in an engraved style in colour on a pale background (see page 12).

Indiennes

These are a version of the Indian hand-painted cottons imported into Europe in the 17th century.

1

1   Viktor & Rolf A/W07. Catwalking.com.

2   Example of a ditsy design by Furphy Simpson.

3   Example of a trompe l'oeil design by Jenny Udale.

4   Example of an all-over design by Jenny Udale.

5   Example of a striped design by Kenzo.

6   Example of a spotted design by Kenzo.

2

Ditsy

Small clustered basic motifs scattered over a background.

3

Trompe l'oeil

A design that looks three-dimensional.

Stripes

Parallel bands of colour.

4

All over

A design that works all over the fabric taking up more space than the ground.

Spots or polka dots

These are round circles of solid colour.

5

6

1

2

3

Ethnic

A design with a foreign or exotic style usually thought of as African or Indian.

Geometrics

Angular designs often abstract or non-representational.

Conversationals

Imagery featuring everyday objects or creatures in repeat and sometimes showing a narrative. Novelty prints on boxer shorts would fall into this category. Often the choice of objects featured can allow for easy identification of the period the fabric originates from, as the objects are fashionable at the time.

Tartan

This is a pattern consisting of criss-crossed horizontal and vertical bands in multiple colours. Blended colours are created where the bands cross each other. This check pattern is known as a sett, which repeats down the fabric. Originated in woven cloth, but now used in many other materials, tartan is particularly associated with Celtic countries, especially Scotland.

Paisley

The paisley design developed from a stylised plant form seen on 17th- and 18th-century Indian cashmere shawls. During the 19th century the town of Paisley in Scotland produced cashmere shawls featuring the design. This pattern has now become synonymous with the name paisley.

4

5

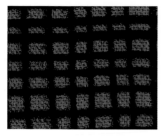

6

7

8

Checks

Horizontal and vertical lines that cross each other at right angles.

Ombré

Gradual shading of one colour into another.

Folkloric

These designs are taken from European cultures and are often representative of a peasant style.

Chinoiserie

French term for European designs based on Chinese motifs, featuring pagodas, dragons, lanterns and temples.

Animal prints

Fake printed animal skins, usually emulating the skins of big cats and snakes.

Psychedelic

Neon acid prints that developed during the 1970s.

Gingham

Horizontal and vertical lines of one colour of the same width that cross each other at right angles producing a small check pattern.

Camouflage

This was invented as a way to blend troops in combat into their surroundings. In the 1970s after the Vietnam War, camouflage military clothing was worn as a rebellious statement by anti-war protesters. Camouflage became more mainstream after being adopted by the youth market.

9

10

1   Examples of an ethnic design by Rory Crichton for Luella S/S04.

2   A conversational design (vintage Liberty).

3   A vintage paisley design.

4   A geometric design by Eley Kishimoto.

5   A tartan design by Kenzo.

6   A check design by Furphy Simpson.

7   A folkloric design by Furphy Simpson.

8   A chinoiserie design by Furphy Simpson.

9   Example of an ombré design by Louise Henriksen.

10 An animal print design by Spijkers en Spijkers.

Print > Embroidery and fabric manipulation

Embroidery and fabric manipulation

Introducing detail to fabric

If detail is required in a design it might be helpful to first draw the design on to the fabric by hand. An embroidery transfer pencil can also be used to trace a design on to paper, the image of which can be then transferred on to the fabric using an iron.

Embroidery can be applied before or after the construction of a garment and concentrated in specific areas or as part of an overall design. It can be used as an embellishment on the surface of the cloth to enhance the look of the fabric or it can be used in a way that makes it integral to the function of the garment, rather than simply as a decorative addition. For example, a buttonhole can be created with interesting stitch work and the shape of a simple garment can change through the application of smocking.

Contemporary embroidery is based on traditional techniques. Hand stitching is the basis of these and once you have learnt the principles, you have the foundation for a vast array of techniques. The three basic embroidery stitches are flat, knotted and linked.

1   A Manish Arora S/S07 design. Catwalking.com.

2   An embroidery sample by Hannah Maughan.

1

Basic stitches

Flat stitches lie on the surface of the
fabric, for example, running, satin
and cross stitch, while knotted
stitches, such as French and pekin,
add texture to a fabric. Linked
stitches are stitches that loop
together, for example, chain stitch.
There is enormous scope for
developing basic stitches. You can
achieve fascinating textures and
patterns by working in different
threads, changing scale and
spacing, working formally, working
freely and combining stitches to
make new ones. Also experiment
with the base cloth you are working
on; the key is to be as creative and
innovative as possible.

2

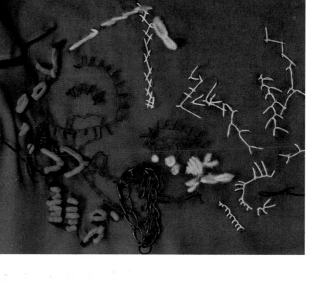

Satin stitch

This is a repeated long horizontal or
diagonal stitch that sits on the
surface of the fabric. The stitches are
worked parallel to each other and
close together to produce a satiny
area where the front and back of the
fabric look the same. The stitch is
widely used in Chinese embroidery.

Cross-stitch

This is often used on a fabric with
an even weave where the threads
can be counted and the stitches can
be exactly placed. Cross-stitch is
often associated with the peasant
look and English Victorian ladies in
the 19th century.

Couching down

This is when threads are laid on the
fabric and small stitches are used to
hold the main thread down. It is a
decorative technique often used
when the main stitch is too heavy to
pass directly through the cloth.

French and pekin knots

The size of the knot depends on the
thickness of the thread used and the
number of times it is wrapped or
looped around the needle to form the
knot. The French knot is wrapped
around while the pekin knot is looped
and neater in appearance. Bullion,
coral and colonial are other examples
of the knotted stitch.

Chain stitch

This stitch can be made with a needle
or a tambour hook. The first looped
stitch is held down with the next
stitch to form a chain.

Blanket stitch

This is used to strengthen the edges
of blankets or garments. Buttonhole
stitch is the same, but worked with
tightly packed stitches for a
stronger finish.

Embroidery techniques

Blackwork

Blackwork became popular in England in the 1500s perhaps due to its popularity with Catherine of Aragon, Henry VIII's wife. It usually features black stitches on a pale background. The stitches are flat and regular in nature creating a graphic effect. Double-running stitch, Holbein stitch and backstitch are usually used.

Assisi style

An Italian style of embroidery where the background of a fabric is worked and filled in with stitches leaving the design motif unworked and in the negative. Originally double-running stitch, Holbein and cross-stitch were used.

Bargello or Florentine work

Working on a canvas, straight vertical stitches are placed in a zigzag design and the colours of the rows of zigzag are changed to create a pattern.

Berlin style

This is an embroidery style (originally from Germany), where bright coloured wool in tent or cross-stitch is worked on canvas.

Canvas work or needlepoint

This is sometimes called tapestry work as the finished stitched piece looks similar to a woven tapestry. The canvas is usually woven using a single- or double- (Penelope canvas) thread construction. Needlepoint work uses a variety of embroidery stitches, but they must be worked closely together so that the canvas is eventually covered and cannot be seen. This is best achieved on an even-weave fabric with the same number of threads in warp and weft, as threads can then be counted and the stitches placed regularly and precisely.

Crewel work

Ornamental needlework, typically using crewel yarn, which is wool of special worsted yarn of two twisted strands.

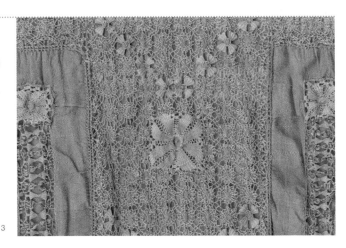

2

Open work

Open work gives the appearance of lace yet is worked on fabric and holes are created through cutting and/or stitch. Examples of open work include pulled thread work, withdrawn thread work, cutwork or eyelet lace.

Drawn work

Warp or weft threads are pulled out of the cloth and the remaining threads are held back with embroidery stitches. The spaces are decorated with stitch work and needlework, which also serves to strengthen the open structures. John Ruskin introduced the technique to linen workers in the English Lake District and Ruskin Lace has been practised since the 1880s.

3

Pulled work

Pulled thread work produces a stronger fabric as no threads are taken away. The lace effect is created through stitches pulling the warp or weft away from the normal weave structure. For the best effect the fabric is loosely woven and the stitches that hold the warp and weft back are of a fine thread so that they don't show. The intricate German derivative of the technique is known as Dresden work.

Needle weaving

This is a variation of drawn thread work. Threads are drawn from the fabric and the remaining warp or weft threads are grouped and woven together.

1   An example of Assisi embroidery can be seen at the top of this sample.

2   An example of crewel work.

3   An example of drawn work and needle weaving.

White work

The open work techniques just discussed are used in white work, but the base cloth and thread stitch are traditionally white. Typical white work also includes broderie anglaise, Richelieu, Dresden and Reticella.

Broderie anglaise

This features rhythmic and repetitive eyelet patterns, where fabric has been cut away and the edges are prevented from fraying by stitches. From 1870 it was produced on a greater scale as it could be done by machine.

1   Example of broderie anglaise.

2–4   Alison Willoughby appliqué skirts.

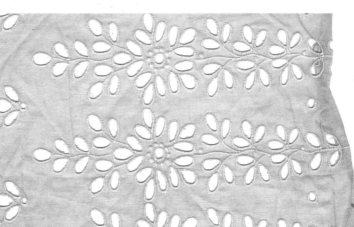

1

Richelieu cutwork

This is French embroidery. Cardinal Richelieu was the principal minister for Louis XIII. He wanted France to be self-sufficient and therefore welcomed Italian lace makers to France to teach their skill. Richelieu cutwork is a development of Venetian lace. Designs feature an organic or floral pattern, with the edges of cut-away shapes defined by stitch, and within the shapes are buttonhole bars.

Mountmellick work

This originated in the town of Mountmellick in Ireland during the 19th century. A soft matt white cotton thread is stitched on to a closely woven white fabric in bold organic designs. Most samples are finished with a knitted fringe.

Shadow work

Stitches are worked on the reverse side of a sheer fabric, usually herringbone or double backstitch. When the fabric is turned over to the right side, shadowy shapes can be seen.

Fabric manipulation

Appliqué

Appliqué in textiles means to stitch one piece of fabric on to another for decorative effect. Pieces of fabric can be stitched on top of a base cloth or a reverse appliqué technique can be applied where the top fabric is cut away to form a pattern and reveal the fabric beneath. Interesting intricate designs can be created using many layers of cloth. Fabrics that do not fray are often good for appliqué work. Fabric motifs, such as badges, can be beaded or embroidered first and then appliquéd on to the garment with stitches.

Smocking

Smocking has a practical as well as decorative function, as it is used to gather in fullness in a garment. Traditionally a garment that featured smocking was called a smock and was worn by agricultural workers in the 19th century. The stitches and motifs used in the smocking related to the trade of the workers.

Horizontal rows of dots are marked on the fabric and running tacking stitches placed at these points. These stitches are then drawn together forming vertical pleats in the fabric. The pleats are then stitched permanently together to create the smocking and the original tacking stitches are taken away. Smocking should be placed parallel to the direction of the warp and weft to avoid distorting the fabric.

Patchwork

The technique of joining together pieces of fabric to make another fabric creates a patchwork; the pieces can be sewn randomly or in a geometric pattern. The choice of fabric and placement of pieces to form the pattern creates the design.

English patchwork
Paper pieces are used as templates to cut fabric shapes; the shapes are cut slightly larger than the template. This extra piece of fabric around the template is then folded over and tacked down. These fabric paper pieces are then sewn together at their edges and the tacking stitches and paper taken away.

American patchwork
Fabric pieces are cut using a template, but the template is then removed. The fabric pieces are subsequently sewn together with a small running stitch. A fabric that allows for a crisp fold works best for patchwork, for example, finely woven cotton.

Quilting

Layers of fabric are stitched together to form a heavier quilted fabric. Wadding of cotton, wool, horsehair or feathers can be put between the layers to make a warmer or more decoratively raised fabric.

English quilting
Two layers of fabric are stitched together with wadding placed in between.

Italian corded quilting
Traditional Italian quilting designs are based upon pairs of parallel lines through which cord or wool is threaded to make a raised pattern.

Trapunto quilting
Like corded quilting, Trapunto is padded after the stitching is complete leaving a design that stands out in relief. Enclosed stitched shapes are slit in the back of the fabric and padded from behind, the slit is then sewn up.

1

2

Machine embroidery

3

4

Many of the embroidery stitches and techniques discussed already can be worked on domestic or industrial embroidery machines. The machines can be used creatively and flexibly to produce a wide range of effects and techniques, from controlled to more freestyle work. Domestic embroidery machines allow the user to move the fabric under the needle in order to create free-flowing designs. Most machines also have the ability to produce automated patterns at the press of a button. As with hand embroidery, the techniques can vary in accordance with the choice of thread and fabric. More complicated embroidery designs can be created on the computer and then downloaded for use on digital embroidery machines. Machines can have single or multi heads to feed many threads simultaneously. Embroidery machines include Cornelli, Irish, tufting, loop-pile cut machines and Schiffli machines.

1–2 Marios Schwab A/W07 garments padded with goose down.

3–4 The white silk fabric of this blouse was first pleated and then embroidered with white cotton. When overdyed the embroidery stayed white while the base cloth coloured blue.

A dissolvable fabric was embroidered with the design in silk. The fabric was then dissolved and the design was stitched on to the organza. Both embroideries were produced on the Schiffli machine; embroidery by Punto Seta for Magdalena Glowacka.

Embellishment

1   A Marios Schwab A/W07 neoprene and metal dress. The metal pieces can be screwed off.

2   A Marios Schwab A/W07 heavily embellished Swarovski crystal dress. Each crystal is enclosed in silk net then hand stitched on.

3-4   Two beaded and sequined pieces designed by Richard Sorger.

Another way to add surface interest to fabric rather than through print or embroidery is to embellish, which gives a more three-dimensional and decorative look. Beads, sequins, mirrors (shisha), seeds, shells, pebbles and feathers can be used to add colour, pattern and essentially surface texture to a fabric or garment. Beads and sequins were used for decorative effect on the flapper dresses of the 1920s, their reflective quality enhanced by the movement of the dresses as the wearer participated in the new dance style of the time. The weight of a fabric can be changed through the addition of embellishments. Beads were used by Fortuny to weigh down the sides of his pleated 'Delphos' dresses (see page 17).

In certain cultures embellishment has been used for social identity or superstitions. Buttons, medallions and braids can show rank and power. Eagle feathers are worn by the Native Americans of North America and signify bravery. Shiny items such as coins or mirrors are commonly sewn on to garments to avert the evil eye.

Consider what kind of embroidery stitch and thread you use to attach the embellishment. Are you able to stitch through the embellishment, as you would with a bead or sequin, or is it held down on to a fabric with a combination of stitches like the application of a mirror disc?

1   2

Beading and sequining

Beads can be made from glass, plastic, wood, bone and enamel, and are available in a variety of shapes and sizes. These include seed beads, bugle beads, sequins, crystals, diamanté and pearls. Beading adds texture to fabric, for example, using glass beads on a garment lends the textile a wonderful, light-reflecting, luxurious quality.

Beads can be stitched individually or they can be couched down, where a thread of beads is laid on top of the surface of a fabric and stitched down with small stitches between the beads. Running the thread over beeswax or a candle before threading beads helps to strengthen the stitch and minimise fraying.

French beading is the application of beads stitched with a needle and thread on the front of fabric that has been stretched over a frame. The frame keeps the correct tension of the fabric, making beading easier and giving the work a more professional finish.

Tambour beading is a technique whereby beads and sequins are applied with a hooked needle and a chain stitch from the back of the fabric. It is a more efficient way to apply beads than French beading.

Types of beads:

Glass beads
These can be transparent, opaque, pearly, metallic, iridescent or silver lined.

Rocailles
These are small glass beads.

Round rocailles
These are smooth on the outside and inside.

Tosca or square rocailles
These are smooth on the outside and square-cut inside to catch the light.

Charlottes
These are ridged or faceted on the outside, square-cut or metal lined inside.

Bugles
These are tube shaped.

Crystal and cut glass
These are highly reflected with many facets.

3

4

Cutwork

Fabrics can also be enhanced through the use of hand cutwork or, more recently, with a laser. Precise patterns can be achieved with laser cutting. The laser also seals, or melts, the edge of man-made fabric with heat, which stops the fabric from fraying. An 'etched' effect can be achieved by varying the depth of the laser cut into the fabric and very complex detailing can be achieved.

Embroidery and fabric manipulation > **Embellishment**

'Colour is an element which right from the initial sketch appears inseparable to me in regard to the idea of a dress, to its substance and nature. My colours are always part of the whole, have a chosen place in the idea from the very first moment of conception.'

Gianfranco Ferre in *Fashion: Great Designers Talking* by Anna Harvey

It is important to understand the principles of colour and how individual colours can be worked together to create palettes that can be used within the design of textile and fashion collections. This chapter looks at the fundamentals of colour, how it is used in design and also its significance to trend prediction.

Fabrics available to a designer are influenced by trends. The colour, fibre and handle of the cloth will most probably have been designed and created based on trend information. Trend forecasters track trends and recognise new directions. Forecasters predict all aspects needed for fashion and textile design such as colour, fibre, fabric, silhouette, details and lifestyle. Forecasters do not have the ability to dictate styles. They forecast when the consumer is ready to accept a new trend and at what market level and price point. This helps designers, manufacturers and retailers select products that are on trend and that the consumer is ready to buy. By anticipating trends, forecasters enable companies to take advantage of new opportunities.

Colour

In order to successfully design a fashion and textile collection colour must be considered. It is fundamental to the feel of a collection and is what the customer first sees. Colour may be chosen for a variety of reasons. It may relate to a season, the profile of a customer, the type of fabric that is available or the concept of the designer. Colour might also be influenced by trend information and a designer may decide to produce a collection that will fit in with the colours predicted for a specific season.

We first need to understand a little of the basics of colour theory and how colours scientifically work together. From this we can start to look at how the designer uses colour within design.

Colour theory

1   The colour wheel.

2   Additive primaries.

3   Subtractive primaries.

4   The CMYK colour system.

Colour originates in light. Sunlight is colourless, but in reality it is made up of colour; this can be seen in a rainbow. Light shines on an object and certain colours are absorbed leaving the remaining colour to be reflected back to the eye. This information is then sent to the brain, which is when we register the colour of the object.

So you could simplify colour down to the colour we can touch, for example, on the surface of an object, like the red of an apple, and the colour that we can't touch, which is made up of beams of light, for example, the colour from a computer screen.

1

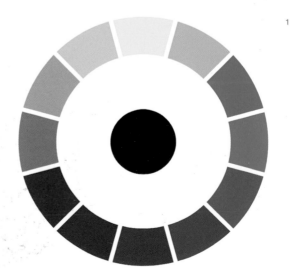

Additive colour system

The basic principle of additive mixing shows that when the primary colours of light – red, green and blue – are mixed in equal amounts they make white. Red and green light produces yellow, blue and green light produces cyan, and red and blue light produces magenta.

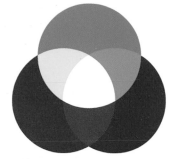

2

Subtractive colour system

Subtractive mixing is a principle where the primary colours of magenta, cyan and yellow can be mixed to produce all other colours. All colours mixed together would produce black rather than white as in the additive system.

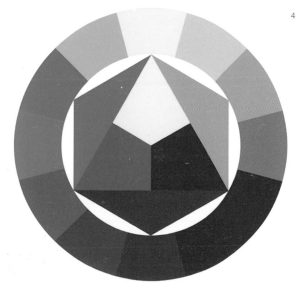

3

The colour wheel

Mixing other hues cannot create the primary colours red, yellow and blue. Secondary colours come from mixing the primary colours together. Blue and yellow becomes green, red and yellow becomes orange, and red and blue becomes violet. Tertiary colours are the colours that come from mixing secondary colours together.

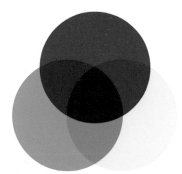

4

CMYK colour system

This is used in the printing industry. Cyan, magenta, yellow and black are the primary colours that make all the others. If you mix all four colours together you produce black.

Darks

1

Brights

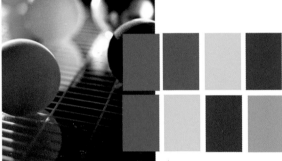

2

Colour definitions

Hue
This is colour.

Saturation
This is the purity of the hue, its richness, strength and intensity. Bright colours are very saturated. Saturation is also known as chroma.

Tone
This is the lightness or darkness of the hue. It is also often referred to as the value of the colour. Dirty colours have more black added to them, pastel colours more white. The tone or saturation of a hue gives a colour many variations.

Fluorescent
These colours react to light and seem to glow.

Highlight colour
This is a small proportion of colour used in contrast to a group of colours to lift a palette.

Tonal colours
These are of the same hue, but range between light and dark colours.

Monochrome
This is a one-hue palette.

Colour harmonies
These are hues that sit well together and have a good balance with each other.

Naturals
These are colours derived from the landscape, sky and water.

Pastels
These are colours lightened with white.

Contrasting colours
These are from opposite sides of the colour wheel and fight against each other. The contrasting colour of red is green, yellow is violet, and blue is orange. The colours that seem to most contrast one another are the primaries as they are the purest colours.

The perception of colour is heightened by the use of contrasts and harmonies within a palette. A red, which is seen as a warm colour, can seem even warmer if it is put with a palette of cool colours. A pure saturated yellow can seem very bright if put with a palette of pale yellows.

Colour palette
A group of colours.

Pastels

3

4

Naturals

The language of colour

It is important to understand the language of colour within design. The human eye can see around 350,000 colours, but cannot remember or recall them all. It is therefore important to have a way to identify and communicate colours. Words are used to describe and give reference to a type of colour by association, for example, pillar-box red or blood red. Words are also used to describe specific colour tones, for example, cool colours have a blue undertone, while warm colours have an orange or red undertone. A washed-out colour could be said to have little hue or to be weak. Pastel colours have white added to them, making them pale, but not weak.

We give colours subjective and symbolic meanings. We apply our own individual characteristics and associations to colour and various cultures see colour differently. In

Europe the colour blue is associated with a boy and pink with a girl, white for a wedding and black for mourning. In India, red is associated with fertility and is also used as a wedding colour, while white is linked to mourning. In most Asian cultures, yellow is the imperial colour and has many of the same cultural associations as purple does in the West. In China, red is symbolic of prosperity, luck and celebration, while white is symbolic of mourning and death.

It used to be that colours were common to a geographical location due to the dyestuffs that came from the minerals and plants found in that region.

Colour psychology

There is a psychology to colour with scientific evidence to show that certain colours affect our mood. There are colours that make us feel depressed and others that raise our spirits; some colours make us feel warm and others cool. Blue is considered to be a calming hue, while black and grey are seen to be depressing. These theories of colour are interesting to consider, but within fashion the choice of colour used within a collection tends to be related to artistic choice rather than psychology. Certain colours are in fashion one season and out the next season regardless of whether they make us feel better or not.

1–4   Colour palettes and image inspiration by Justine Fox. Copyright Global Color Research Ltd.

Colour > Trend prediction

Colour and design

The choice of colour within design is quite a personal thing. We all have our own personal palettes that we like to work with – colours that we feel are exciting, comfortable, classy or fun. As a designer you may have to work outside your own range of colours with palettes you are not very comfortable with. It is therefore important to try and understand how colours work together, and experiment.

Certain designers are known for their use of colour. The Japanese designers Comme des Garçons and Yohji Yamamoto tend to use dark colours. Their collections are timeless and concentrate more on the clever cutting of a garment than a fanciful colour. Versace, however, relishes colourful collections to seduce its customer. Marni and Dries Van Noten use colour beautifully, their palettes are sophisticated and unusual. Calvin Klein is known for its muted neutral tones and Tommy Hilfiger for bold primary colours.

Certain colours such as red, navy, black, white and ivory are so basic they are always fashionable for mass-market end usage. Menswear colours tend to use these safer colours in mass-market and high-end fashion.

1   Dries Van Noten A/W07 runway show. Catwalking.com.

2   Backstage at Louis Vuitton's S/S08 show.

Texture

Of course, within fashion a colour does not work on its own. The designer will see the colour in relation to a surface or textile, and in the context of a silhouette or garment, and this can change the perception of the colour. For example, the quality of a colour can change in relation to certain fabrics – red can look cheap and playful in a plastic, but it can look luxurious and rich in a fine silk. Black polyester can look cheap, while black wool can look very expensive (obviously this also depends on the quality of the fabric of choice). Lighter colours show texture better than darker colours.

Proportion

It is important also to consider the proportion of colour within an outfit. Sometimes difficult or unusual colours are best dealt with in smaller proportions, but it all depends on the customer and trends in colour at the time. A new fashion colour (one that has not been in fashion before) may be first introduced in small amounts within a print or multicolour knit, or used as an accent (highlight) within a group of colours.

The placement of a colour on the body can make certain areas look bigger or smaller. Black is seen to recede to the eye so making an object seem smaller; this principle can be used to flatter the body shape.

Context

It is also important to consider the context in which colour is used and what it is trying to communicate. For example, in the West a red wedding dress is conveying a very different statement to a traditional white dress. Also consider how colour has been used historically for certain garments, for example, indigo denim jeans, the white shirt and the little black dress. If the colour of these staple garments is changed, do they then become faddy and not classic?

Colour can help to keep product lines new and fresh. Often a garment does not change each season in silhouette or detail, but it does change in colour.

2

Colour > Trend prediction

Khaki

During their years of colonial rule in India, The British Army dyed their white summer tunics to a dull brownish-yellow colour for camouflage in combat. This neutral tone was called 'khaki'. The word's origin is mid-19th century from the Urdu term *kaki* meaning 'dust-coloured' and from the Persian word *kak*, meaning 'dust'.

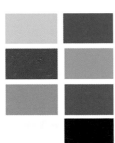

1

Season

Colours can also be seasonal. Cold seasons tend to warrant darker colours, such as blacks, browns and sludgy colours. As the season warms up the colours become lighter and paler. They then become stronger and brighter as the sun becomes more intense. The sun bleaches out pale colours, so if you are designing for hot countries consider a brighter colour palette. Think of the colour palettes of African textiles or Hawaiian shirts. When we pack for our summer holidays we quite often take brighter clothes than we would wear in a colder climate.

2

Colour referencing

Colour often needs to be consistent across various fibres or fabric types, which in turn may require different types of dye that may even be produced in different countries. For a colour of a textile to remain consistent from the design stage through development to realisation, companies often use a colour referencing system. Pantone and the Munsell colour systems are common references for colour matching, as each colour has a specific number for reference. Rather than trying to describe the colour, the number can be used to identify the hue. Pantone charts are arranged chromatically by colour family and contain 1,925 colours. They are a great resource, but they are expensive and need to be replaced as the colours start to fade, making referencing inaccurate.

Looking at colour under different lighting conditions can affect the hue – an incandescent light places a yellow cast on the hue, while a halogen light creates a blue cast.

Colour and the customer

Colour is very important within fashion and textile design. When a customer enters a store they tend to be drawn to the colour of a garment. They may then go and touch the garment and lastly they will try it on to see if the fit is right.

Within a fashion collection safe colours are usually black, navy, white, stone and khaki. Buyers will often buy in garments in these colours as they are the staple colours of most people's wardrobes. It is sometimes a good idea to offer some of the basic colours and add to them seasonal experimental colours. These colours will add life to the collection and will ideally entice the customer to buy each season's new colours along with the trans-seasonal basics.

Skin tone can also have an effect on the colour choice of a garment. Dark skin looks great against strong, bright colours, while softer colours work better against paler skin.

1   A colour palette created by Justine Fox in response to the Chloé S/S08 collection. Copyright Global Color Research Ltd.

2   Chloé S/S08 runway show. Catwalking.com.

3   Pantone colour book.

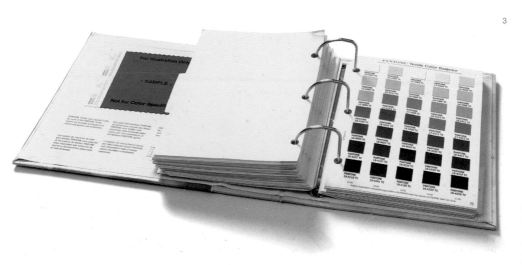

3

Trend prediction

1   Mens- and womenswear
     trend information. These
     pages show inspirational
     images, colour referencing
     and suggestions for colour
     groupings and proportions.
     Copyright Global Color
     Research Ltd.

Trend prediction is a huge industry influencing many areas of design especially fashion and textiles. A trend company focuses on a specific area of a market and tries to predict what is going to happen to that market sector in the future. It could be that the trend company is predicting what will happen next season or what may happen in years to come. These companies sell the information they gather to other companies who do not have the time or the resources to do their own research and prediction. More often this information is sold to companies that do some form of trend prediction in-house already. They then align their trends to those of the prediction company in order to feel comfortable that what they are designing is on trend, that the market is ready for the designs, and that it understands them and ultimately wants to buy them.

Trend companies look at the progression of trends. They monitor what has been successful for a while and evaluate whether this trend will continue and grow stronger or whether it is time to react against it and do something different. Trend companies look at a variety of sources to inform them. They may look at what is happening in society, the economy, the arts, fashion, science, street culture and haute couture, for example. It is easier to identify a trend once it has happened; looking back, historically important trends can be recognised through changes in fashion. Trends are far clearer in retrospect.

Certain design companies are keen not to follow trends and instead be seen to be setting the trends. These companies work with their own prediction ideas and in a way operate like fine artists, developing their own personal ideas and concepts. These designers have to find their own niche market that is not influenced by trends, which can be difficult. The basis of trend prediction is intuition, what feels right and what feels new.

1

FLASH
nocturnal intrigue

AW 07/08

AW 07/08

MIX FASHION WOMENSWEAR
ISSUE THIRTEEN

FLASH 13

Crystal paving   Pompom trims   Metaloplastic weaves   Glitter paints   Dimensional jewel appliques

FLASH
electric

AW 07/08

AW 07/08

MIX FASHION MENSWEAR
ISSUE THIRTEEN

FLASH 13

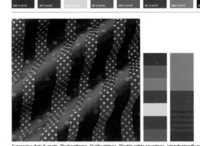

Expressive dots & spots   Pixel patterns   Quirky stripes   Electric cable coverings   Unashamedly man-made

Colour > Trend prediction

Trend sources

Trend companies see that there are long-term trends and short-term trends. Long-term trends look at social trends, demographics, global trends, new technologies and processes. For example, an increase in the use of the Internet facilitates easier communication and enables employees to work more from home, so this might have an influence on fashion becoming more casual and comfortable. Short-term trends are more affected by passing fads, for example, an important retrospective exhibition or a hot new designer's current collection.

1

Culture

Until recently fashion was for the wealthy upper classes and nobility. The lower classes looked at what they were wearing and emulated it (known as the trickle-down theory). This, however, changed in the 20th century as street fashion started to trickle up and be adopted and reinterpreted by the couturiers. Now fashion trends are seen to trickle up and down, influencing consumers up and down the scale. This in turn drives new trends. As street fashions become too mainstream and popular they are seen as unfashionable with the style setters. As a result new styles emerge.

New technologies

New technologies and processes lead to new developments within the fashion industry. These can be in the form of new fibres, yarns, printing processes, dyes or manufacturing processes, which in turn trigger new colours or fashion silhouettes, creating new trends. The copper roller-print process allowed for lengths of printed fabric to be produced first in one colour then in multiple colours. These printed fabrics can be seen in the fashions of the late 18th and early 19th centuries. More recently the development of the circular knitting machine has allowed for seamless underwear and also Issey Miyake's A-POC concept. New developments in nano-fibres are creating exciting possibilities for interactive textiles.

1–2   Scenes from the Première
       Vision Paris trade show.

2

Shop reports

Shop reports analyse what is happening in store at a particular time, in other words, what the shops are buying and putting in store and what customers are buying. Looking at the best stores in a city can give a good overview of the strong fashion trends for the season. It is often a good way to see the collections from up-and-coming designers in high-end boutiques. These designers may have new innovative ideas, but cannot yet afford to show their collection on the catwalk and receive press coverage from their shows.

Trade fairs and the catwalk

Communication is now so quick that trends are disseminated around the world in seconds via the Internet. A catwalk show in New York can be seen an hour later in London and used as inspiration for a high street fashion company immediately. Companies are able to react to this information directly and produce collections to go in store within weeks. Trends now travel and are picked up faster than ever, therefore new trends are replaced far quicker than before.

Fabric fairs such as Première Vision and Expofil (France) and Pitti Filati (Italy) all feature trend areas. Here a presentation is shown that highlights the predicted textures, yarns and colours for the season ahead. It includes fabrics from the companies exhibiting and colour palettes that can be purchased with colour referencing for exact colour matching.

Intuition

Within trend prediction nothing is fact, all information is up for interpretation and reinterpretation. However, it is clear that certain individuals just have a knack for interpreting information and successfully predicting trends. Natural intuition has a great role to play in trend prediction.

Flow of information

There are levels of trend prediction within fashion. The first level looks at trends in colour. Colour groups meet from around the world to put forward their ideas for future colour palettes. These predictions can sometimes occur two years before the collections are seen in store. It is important that the chemical companies that produce dyes know what the colour trends are going to be so that they can supply appropriate dyes to the fibre, cloth and garment dye industries. It can take dye manufacturers four to nine months to manufacture dyes and send them to the dye houses. The darkness or lightness of a colour on trend will affect the amount of dyestuff

that is needed, also the kind of fabrics that are going to be dyed affects the type of dye required. In the 1950s ICI produced new, cheap, bright dyes called Procions for cellulose fibres, which were seen in the brightly coloured fashion of the 1960s.

The second level is texture and fabrication, determining the fabrics that are going to be important. Fibre companies, mills and weavers will look to this type of trend prediction to see what new fibres have been invented or improved upon. They look at different mixes of fibre types and how they perform, and also at the prediction for yarn weights and

finishes. Large fabric companies such as DuPont suggest colour and tactile qualities.

The third level is surface interest, in other words, the print and embellishment for the season, for example, strong print ideas and key motifs and colours. The fourth and final level is the garment trend prediction, in other words, the key garments for the season and their details and silhouette.

Cool hunters

Cool hunters are employed around the world to find what is new and 'cool'. They usually look at underground events and movements that are not known about in the wider community, but are strong ideas within a core niche of society. They look for things such as a new band, a new store, a new toy or a new way of wearing trainers or jeans.

Colour groups

Most countries have groups that brainstorm colour trends for the home market and for export. In the United States there is the Color Association and the Color Marketing Group, in Europe the Deutsche Mode Institut and the Institute da Moda Espagnol. Colour groups are made up of the leading fashion colourists from fibre companies, fashion services, retailers and textile firms who develop and produce colour palettes for fashion and furnishings. The palette can include a large number of colours and it is important to look at the hue, value and intensity in relation to its usage on a product range.

1   A selection of interiors from a range of Marni stores.

1

Colour > Trend prediction

1

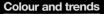
Packages and presentations

Packages are produced by the trend companies to sell to the fashion industry, which contain information on the specific trend areas. They may contain colour palettes with reference to Pantone colours; images are grouped to form moods for the different stories or ideas and fabric swatches and yarns may also be included. Sometimes the fabric swatches can be very experimental, developed specifically by designers working on the new trends.

Images of catwalk shots or fashion illustrations help to describe future garment trends.

The trend packages can be quite specifically targeted to a design team's needs and not shown to any other designers or they can be more general, targeting more companies. Obviously the more selective the package, the more expensive it will be.

Some trend companies are involved in presentations where a member of their team talks through a series of projected images and the images are often presented with key words to stimulate the imagination.

The main trend companies are Trend Union (Europe), Nelly Rodi and Peclers (France), INDEX and Global Colour Research Ltd (UK) and BrainReserve (USA). The two trend 'gurus' are Li Edelkoort of Trend Union and Faith Popcorn of BrainReserve. They are most successful at being able to read the clues that indicate trends that are already out there or up and coming.

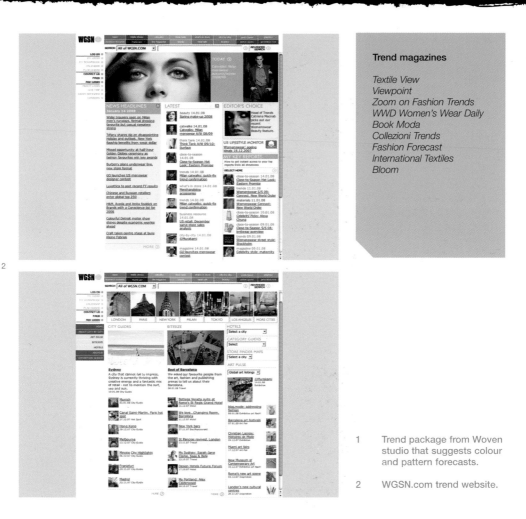

2

Trend magazines

Textile View
Viewpoint
Zoom on Fashion Trends
WWD Women's Wear Daily
Book Moda
Collezioni Trends
Fashion Forecast
International Textiles
Bloom

1   Trend package from Woven
     studio that suggests colour
     and pattern forecasts.

2   WGSN.com trend website.

Publications and the Internet

There are various trend magazines
available that include exciting
inspirational ideas. These magazines
commission images that sum up
a trend idea. They may also
commission fashion illustrations to
depict key fashion silhouettes and
details. Colour palettes are also
featured and there may be a round-
up of the season's catwalk pictures.
Some publications are very inspiring
and allow the reader to really
interpret the ideas suggested,
others are more commercial and
defined in the ideas they publish.

The Internet is a great place to find
trend information as it can be
updated so quickly. Millions of other
people can also get the same
information, which means a trend
can spread quickly, however, the
downside is that a trend may as a
result suffer from overexposure.
Many trend websites are
subscription-only and as a result,
information remains exclusive to the
subscribers. Worth Global Style
Network (WGSN), established in the
UK in 1998, has offices in the major
capitals of the world and is seen as

the global online trend leader. The
key fashion and textile companies
subscribe to this service and are
given global news, reviews and
inspiration. WGSN offers a free
subscription to students, however,
the information they receive is a
couple of weeks old.

Trend prediction is a useful tool,
but should be used creatively
and interpreted with
independent thought.

Colour > **Trend prediction**

'I build my designs from a lot of small pieces which I attach to each other in different ways to discover the shape that I want. In that sense I guess you can say that I approach fashion more as a sculptor than a tailor.'

Sandra Backlund

This chapter focuses specifically on how textiles are used within fashion. It investigates the decisions a designer has to make when choosing a fabric, choices to do with functionality, aesthetics and cost. It also looks at how a designer can best design with fabric, working in three dimensions by draping on the stand or through computer draping. When the textiles have been designed and made it is important to understand how to then make them into garments, what are the best ways of cutting and constructing specific fabrics. The second part of this chapter gives an enlightening insight into how textile designers work in the fashion industry through interviews with designers working as freelance textile designers, fashion textile designers, designers in the fields of trend prediction and designers who are pushing the boundaries of future textiles.

Choosing fabrics for fashion design

1   Jessie Lecomte A/W07
     wool coat.

As a knitted textile designer it is possible to design just the knitted sample, but you are more likely to design the sample and also the garment that the sample becomes. These two processes are very much connected, as the garment literally grows from the knitted stitch.

As a print or weave designer the process of textile design and garment design are less integral. The fabric tends to first be designed as a length and then the garment is cut from it. However, some of the most interesting print and weave designs can come from the knitwear approach where the garment shape develops along with the fabric sample so that they connect with each other.

As a fashion designer it is extremely important that you understand what properties fabrics have and how best to use them on the body, functionally and aesthetically.

The best fashion designers have a strong understanding of fabrics, how best to design with them and construct garments from them. The best design ideas on paper will remain just drawings unless the designer has these skills. Try to integrate the design of the silhouette and details with the choice of fabric as you go along. Fabrics can stimulate garment ideas and vice versa. Certain designers will be known for their use of textiles for fashion, others for their details and silhouettes, but these designers still need to choose the right fabric for their designs. A poor design can be improved with a fabulous fabric, but a fabulous design rarely works in a dreadful fabric.

Choosing fabrics for fashion design > Designing with textiles

1–4   Jan Taminiau A/W07. In this collection each garment can be worn in two ways, so every garment came out twice on the catwalk.

1–2   The jute base cloth of this dress is tufted with silk chiffon and silk crepe. The top layer is made of washed silk and tapespooled lace woven with needle-spooled lace.

3–4   This one-piece woven dress is constructed from three layers of cotton, with elastic woven into it to create shape. Underneath is a boxer short.

Consider what you are trying to achieve in your fashion designs. Are you after a flashy print or embellished garment to dazzle as a showpiece on the catwalk? You may need a fabric that will show off design details of cut, seam lines and darts – a woolly knit certainly would not allow for this, but a simple plain weave would work well. What kind of silhouette are you working with? A fitted silhouette close to the body can be created with a tailored woven fabric or may be a stretch fabric or bias-cut fabric. A silhouette that sits away from the body could be created from thick boiled wool or perhaps from a crisp organza with French seams giving structure to the garment.

The season you are designing for can dictate the choice of fabric. Heavier fabrics are used more in autumn/winter and lighter, breathable fabrics in the spring/summer. However, we tend now to be able to wear a variety of fabrics in all seasons as we live and work in heated and air-conditioned environments.

Consider the durability and function of the fabric, does it need to be hard wearing or wash well for everyday use or is it a garment that will be worn on special occasions and dry-cleaned? If the garment needs to be worn in poor weather conditions consider the fabric and construction finishes best suited for this.

Lastly, how much does the fabric cost that you are using? Is it appropriate to the level of the market you are targeting? A couture garment will feature the most high-quality original fabrics available. High street designs will be made from cheaper, high-performing fabrics that are durable and wash well.

1

2

3

4

Choosing fabrics for fashion design > Designing with textiles

Function

Performance

As discussed in the previous chapters, performance or technical fabrics can be created at various stages of production, at fibre production, garment construction or even when the garment is finished. Performance fabrics can be creatively used in garment design. A viscose microfibre could incorporate microcapsules containing specific chemicals that when made into a yarn and woven, produce a fabric with UV protection, which could be used for beachwear or childrenswear. Wool could be boiled and quilted to protect the body from the cold or even from pressure or abrasion, ideal for outerwear. A cotton fabric could be waterproofed with the addition of a laminate finish for sportswear. New developments in smart fabrics can also add more futuristic properties to fabrics, for example, fabrics that have a memory can change colour or can even act as a communications interface.Designers can use technical performance fabrics for their aesthetic qualities rather than for their primary functions, for example, neoprene is used for diving suits, but can also be used to create structure due to its density.

Drape

Fine fabrics that have a loose construction tend to drape better than thicker fabrics with a tighter construction. However, this is not always the case, fibre content and finishes both play a part in the drapability of a fabric. It also depends on what type of draping you require; flat fluid drapes or full voluminous folds and shapes. When buying a fabric, unwind a length and hold it up to the body to see how it falls and if it is appropriate for draping.

Volume

Volume can be achieved through the use of thick or hairy fabrics, but also by using large amounts of thinner fabrics that can be gathered or pleated. Fabrics can be used in garments in such a way that they catch air when the garment is in motion, creating volume. Volume can also be created with the use of seams and darts to create shape. It is important to think where the garment is touching the body and what shapes you are trying to make between the body and garment.

1   Alexander McQueen
     A/W06.
     Catwalking.com.

2   Knits from Sandra
     Backlund Ink Blot Test
     collection.

1

2

Structure

Structure can be achieved by employing tailoring techniques adopted from menswear design through the use of specific seaming methods, interfacing, canvases, padding and boning. Structure can be simply achieved by using appropriate fabrics; on the whole a tightly or densely constructed fabric will offer greater garment structure than a loosely constructed fabric. A structured garment does not rely on the body to give it shape, you can create a shape that sits away from the body or that exaggerates the body in some way, for example, a tailored suit could give the illusion of broad shoulders. Structure can also be used to control the body and force it into new shapes. Tailoring was used before the invention of stretch fabric to create new body shapes through clothing.

Stretch

Stretch fabric allows garments to fit on the body without the use of tailoring. Stretch garments are comfortable and easy to wear, allowing the body to move freely with the garment. They also support the body (powernet can be used to lift and hold the body) and if used well stretch can be used to flatter the body. Stretch fibres are frequently added to other fibres to improve the performance of a fabric. Fabrics tend to distort when we wear them especially in areas such as the knees, elbows and seat. A proportion of stretch in a fabric will bring the fabric back into shape after it has been worn.

1   Christopher Kane S/S07. Catwalking.com.

2   Examples of fabric selection and use within the design process.

3   Tata-Naka hand-knitted jumper with three dimensional parts.

1

Aesthetics

Colour, mood and trend

The choice of textiles, imagery and colour significantly influences the message of a fashion collection. As discussed in Chapter five: Colour and trends, most fabrics that are produced have been designed as a result of the influence of trends. These trend ideas might then come through into a fashion collection or the fabric may be used in a completely different way. For example, a futuristic laminated fabric could be used to produce a modern, futuristic-looking fashion garment or the fabric could be used to give a modern twist to a classic piece. The use of well-chosen fabrics can unite fashion ideas into a strong collection.

Pattern

Pattern can be used to give a fashion collection a specific look, but this can have both positive and negative outcomes. For example, if a designer uses a certain pattern that is not on trend for the season, the collection may not be desirable and may not sell even if the garment shapes are good. Good use of pattern and clever placement can create a very personal fashion collection.

Pattern can be used to create a strong brand image too: think of Pucci prints or Missoni knitwear. However, if the pattern becomes too popular it can have a negative effect on the brand.

Texture

Texture can enhance a garment through its visual and tactile qualities. It is really important how a fabric feels; fabric is worn next to the skin and is felt throughout the day. Texture can also add interest to a garment without using detail.

2

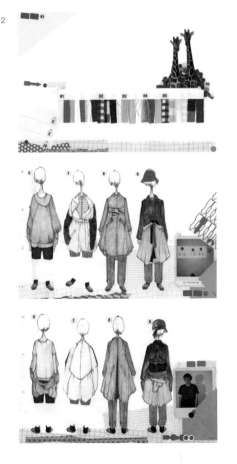

3

Cost and availability

You must consider where you are going to buy your fabrics from if you are not creating them yourself. You may just need a length of fabric for a one-off garment or you might need to buy more of the fabric if you are going to put your fashion designs into production. If you are creating one-off garments you could find fabric at markets, vintage fabric fairs, flea markets or on trips and holidays. If you sell quantities of your designs you will have to think about buying fabric from a place that can supply larger amounts and that you could go back to and reorder from if necessary. If you are buying fabric in a shop you can see how much is on the roll and hope that it is still there if you need to come back. It is safer to buy from wholesale fabric suppliers who will have specific fabrics in stock, as they have catalogues that you can buy from and reorder. However, this can still be risky as they may not have fabric in a certain colour in stock when you need it.

Looking for fabrics at fabric fairs such as Première Vision is an option, however, for students this may simply not be realistic as fabric suppliers must sell fabrics in minimum lengths and you may not be able to meet their requirements. Also certain fabrics that are shown at trade fairs do not go into production if the supplier does not get enough orders for them. When buying from fabric fairs, it is important to check the prices carefully and to find out whether there are hidden costs, such as delivery or supplementary (such as minimum order) fees. Many suppliers will also require a VAT number.

Show pieces
These garments never get to the shop rail, but are conceived to attract press interest, which will promote the designer to a wider audience. They are intended to grab attention.

1

Market level and genre

Consider what level you are designing for. The fashion industry can be simplified into the following categories: supermarket, high street, independent designer (producing smaller amounts of garments), ready-to-wear designers (that show at the fashion weeks in the main capitals of the world), luxury super brands (such as Gucci and Prada) and couture. There are also casual and sportswear brands that range from small labels to the massive super brands such as Levi's and Adidas. It is interesting that most fabrics can be found at all levels of the fashion industry; what matters is the type of garment that the fabric is made into and its perceived value (that is, what the customer expects to pay for it). However, there are some fabrics

that you will see more of at certain levels, for example, sportswear and casual wear will use more technical performance fabrics with durability and stretch. Supermarkets will use cheaper fabrics, but because they will be producing huge quantities of garments they will need a lot of fabric, so can therefore buy a good fabric at a cheaper price than an independent fashion designer. High street garments at mid prices should wash and wear reasonably well; this level of the industry is very competitive and the customer will demand that garments have these qualities. The ready-to-wear level will try to use innovative individual fabrics to set garments apart from other ready-to-wear designers. Showpieces for ready-to-wear designers may not have to perform

beyond the catwalk. These textiles might never need to be washed or need to be very durable, allowing room for more experimentation. Also the work required to create the textile may make it unfeasibly expensive to produce to order. Couture is really the only area where a fabric could be very expensive and it may not need to be washed or wear well.

Don't forget to consider whether you are designing menswear, womenswear or childrenswear. Certain fabrics might be difficult to use in menswear as they may appear too feminine. There are safety regulations about certain childrenswear fabrics too, especially nightwear.

1   Dior Couture S/S07. Each layer of the dress has been hand dyed. Catwalking.com.

2   Prada menswear A/W07. Catwalking.com.

2

Designing with textiles

1   Draping on the stand.

2   Fashion illustrations
        by Julia Krupp.

3–4   Examples of textile use
        within fashion design
        development.

The best fashion collections integrate fabric design and fabric selection with garment design from the start. It is important to integrate the fabric and garment together working from one to the other. To achieve this, select fabrics as you design garments and continue to perfect your fabric choices as your garments develop. It is important to handle (feel and drape) your fabrics when you design in order to understand their properties, for example, whether they drape and stretch or whether they are stiff and structured.

Three-dimensional work

Draping on the stand

1

Certain fashion ideas are best designed in three dimensions on the stand, as working in this way allows you to see how fabrics drape or fold. If you can use the fabric you have designed to drape, this will obviously give the best results, however, you may not have yet created your fabric. If so, choose a fabric with similar characteristics in weight and construction to your designed fabric in order to get as true a representation of the fabric's qualities as possible.

When draping, design ideas are endless so consider what you are trying to achieve, what is the reason for the draping. Drape with control, don't just scrunch. Think about where the fabric touches the body and the shapes you are making. Remember that draping on the stand isn't just about letting fabric fold and hang, it can also help with other areas of design such as the proportion of detail on the form, volume and placement of pattern. It is really important as a textile designer who works in fashion, to trial your designs on the stand or body. Also hold your fabrics up and look at them in a mirror to see if colour, pattern, proportion and texture are working well.

Photograph or draw your stand work and then work into the images adding details, eliminating areas that don't work or changing proportions. Once you have created something you like on the stand, consider how the shapes and proportions you have created work with the body, how they will become garments; in other words, how you will get into them, seam them and finish them.

Digital drape

There are computer packages that allow a designer to digitally drape fabrics on a three-dimensional form. By scanning in a fabric or using a virtual fabric construction, drape, colour and texture can be shown in three dimensions on the two-dimensional computer screen. The benefit of this system is that sample garments can be trialled in virtual space quicker and more cheaply than in reality. The garments do, however, look rather computer animated and designers do not really get a good understanding of a fabric unless they are handling and interacting with it.

Two-dimensional rendering

Drawing textiles

Experiment with different media and textures when you draw garments and think about using a technique that can represent a fabric well. It is important that you can express the type of fabric a garment is made from and not make everything look like it is made of cardboard. Try to express structured, hairy, woolly, flat, smooth, transparent, shiny, hard, soft, padded, crispy, lacy, printed, embroidered and sequined textures. Explore how a fabric moves, drapes and folds on the body.

2

3

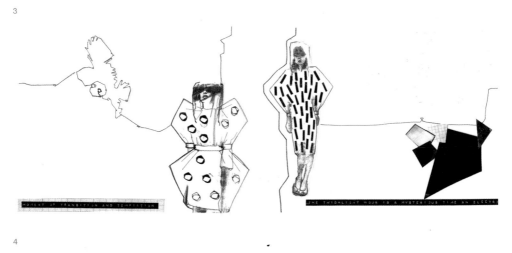

4

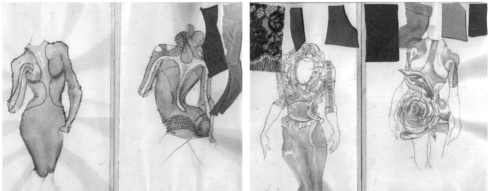

Garment construction

Most fabrics have a front (or 'right') and back (or 'wrong') side, the front being the side that is usually cut to be visible on the outside of the garment. Some fabrics also have a top and a bottom; they might have a repeated direction of pattern like a printed, damask or brocaded fabric. A fabric may also have a pile that has a slight direction and a slight colour difference that can be seen when the fabric is draped, for example, velvet and corduroy. It is important to consider the right, wrong, top and bottom of a fabric when cutting out pattern pieces. Grain lines with a direction should all face the same way on the fabric to avoid cutting a piece wrong.

Garments are normally cut with the major seams running parallel to the lengthwise grain as this helps to control the structure of the garment and there is also more stretch across a fabric. Pattern pieces for sleeves and legs where elbows and knees flex will be accommodated in the stretchy part of the fabric. The bias is at 45 degrees to the warp or weft. Garments can be cut on the 'bias' or cross, which gives characteristic drape and elasticity to a garment.

Selvedge
This is the edge of the fabric running down the length or warp produced during manufacture so that it does not fray.

1   Vintage Comme des Garçons deconstructed wool coat.

2   Detail of a welt seam with top stitching, often used for denim garments.

3   Overlocked jersey seam and bound neckline.

4   French seam detail.

5   Running seam with overlocked edges.

1

Seams and darts are needed to render a two-dimensional fabric into a three-dimensional garment. A seam can be chosen for its functional or aesthetic qualities. Garment seams need to be finished to stop the seam allowance fraying and to make the inside of the garment more attractive. Some seams can be bound or overlocked after construction, while other seams such as French seams and welt seams are constructed and finished at the same time.

It is important to know how best to cut seams and finish garments using the fabrics you have chosen. Good construction will make a garment far more successful and show your fabrics off well.

Seam allowance
This is the amount added to the outside of the pattern edge to allow for sewing.

2

3

4

5

Toile

Toile is a French term for cloth, but is used nowadays to describe a mock-up of a garment (to check fit and make). A toile is usually made from cotton calico (in various weights) or a fabric similar to that which the actual garment will be constructed from. Toiling your designs is important as it allows you to see whether your design ideas actually work as a garment.

Before you make up a garment, measure a square of the fabric, then wash and remeasure it to see how much the fabric has shrunk. When lining a garment make sure the lining fabric has the same wash qualities as the main body of the garment, as shrinkage can cause the garment to hang badly or the colour in a fabric might run.

Certain construction techniques need a fabric that can be shrunk with steam, such as a wool used in a tailored sleeve head. A synthetic fabric will not work in the same way so may not be suitable for a particular technique.

Don't let stretch fabrics hang over the table as you cut them, as the fabric will distort and affect how the pattern pieces are cut. Use stretch interfacings with stretch fabrics if you need to retain the stretch in a garment.

Thin needles, for example, size 9, are used for light fabrics including silk, chiffon and voile, while a heavier needle, for example, size 18 is used for denims, canvas and overcoating. Replace needles as they become blunt for better stitching.

1   Internal construction of
    a La Petite S***** wool
    dress. It is lined with silk
    organza, the hem of which
    is babylocked and shows
    beneath the dress. A
    corseted structure holds
    the top of the dress,
    leaving the wool to hang
    from under the bust.

2   Wool dress by La Petite
    S***** A/W07.

2

Patterned and embellished fabrics

Patterned fabrics

Consider how pattern is placed on the garment. Some patterns may have a direction – a top and a bottom. How does the size of the pattern relate to the size of the garment? Large-scale patterns will need thoughtful placement; a larger garment will obviously provide more space for a larger design. You may need more fabric than you think for your garment so that you can place patterns in specific areas. It is possible to engineer designs so they flow from one pattern piece to another or think about shifting seams so that a pattern can work around the body and not be cut by a seam. If the pattern needs to be symmetrical across a garment, you must take time to carefully lay out pattern pieces on the fabric. When marking out pattern pieces for cutting make sure you consider where the seams join and not where the edges of the seam allowance join. Also look at how a pattern travels across the body and across openings and the centre front of a garment. Consider what the garment looks like as pattern pieces go off grain, for example, a batwing sleeve placed on a vertically striped fabric will have stripes running up the body, but horizontally across the width of the sleeve. It might help to draw the pattern you are working with on to the toile to work out its position on the body and across seams.

Sequined and beaded fabrics

Sequined and beaded fabrics may need to be stabilised before they are cut so that the beads do not come off. The most professional way to work with embellished fabric is to first mark the pattern pieces on the fabric, then remove the sequins or beads from the seam allowance just past the seam line and knot off the bead or sequin threads. The fabric can then be cut without the sequins or beads falling off. Tape can be used to push the sequins or beads away from the seams before sewing and then sequins or beads can be sewn by hand over the seams on the outside of the garment. If a sequined garment is not constructed in this way the garment might have to be lined to cover any scratchy seams that contain cut sequins or beads. Garments should be stored flat to avoid sagging caused by the weight of the embellishments.

Pleated fabrics

Natural fibres can shrink when a fabric is pleated so pre-shrink the fabric first. Flat pleats are best hemmed before pleating, whereas raised pleats can be hemmed afterwards.

1   Pleated double sequined dress by Bérubé A/W07.

2   Jessie Lecomte A/W07 creation.

3   Giles Deacon A/W07 hand knitted oversized cardigan by Sid Bryan.

Knit

Knitwear can be constructed in three different ways. First, fabric can be knitted as a length, then the garment pieces cut and sewn together. Second, garment pieces are knitted to shape or fully fashioned, then sewn together to produce a garment. Finally, the garment is knitted in three dimensions with little or no seams (see Chapter three: Fabric construction for more information on types of knit).

2   3

Cut and sew

Knits have different stretch qualities related to the stitch construction and fibre content, and knits can stretch across length and width or in all directions. Consider where you require the stretch on a garment, usually where elbow and knee joints are and in the seat of a garment. Consider how much stretch is needed to get a garment over your head without a fastening.

Machine-knitted fabrics tend to twist in production so the grain is off; this can cause garments to hang badly. Follow a vertical rib or wale to determine the grain of the knit and lay your pattern pieces accordingly. Don't let the fabric hang off the end of the table as you cut it as this will cause distortions.

Silk jersey can be pinned first to tissue paper, pattern pieces can then be laid on and cut out with sharp scissors. Certain knits can ladder and will curl when cut (single jersey curls to the right side when cut). A ballpoint needle or stretch needle will glide between the yarns rather than piercing the fabric and causing laddering.

Jersey garment seams are overlocked together, this stitch allows the seams to stretch with the garment and not break. The stitch also contains the raw edges of the pattern pieces to create a neat finish. If no stretch is needed in a seam it may be advisable to stabilise it with a non-stretch tape, for example, on a shoulder seam. Reinforce buttonholes with woven interfacing and hand stitch zippers in place before machine stitching them. Let the garment hang overnight before measuring and hemming, then zigzag or twin needle the hem. Knitted garments should ideally be stored flat so they do not sag.

Kenneform

MADE IN ENGLAND
BY
Kennett & Lindsell Ltd
ROMFORD ESSEX
KENNETT & LINDSELL LTD

1

Fully fashioned knit

Knitted fabric that is not cut into pattern pieces must instead be fashioned or shaped into pattern pieces through increasing and decreasing stitches. Fully fashioned shaping can also create a decorative feature at seams. Slight increases and decreases can be created by changing the stitch tension row by row, by changing the thickness of the yarn or by changing the type of stitch. A ribbed edge is a good example of this; ribbed panels can be placed at the waist to bring the garment in to fit.

In order to work out a pattern for a knitted garment you must first knit a tension swatch. Knit a 15cm square of knitting, finish it in the way the garment would be finished, for example, washed or pinned out and steamed, but be careful not to overstretch the sample while you do this. When the sample is dry and relaxed measure the number of rows (vertically) and stitches (horizontally) in a 10cm square section in the middle of the swatch. In certain swatches you may have hidden some of the stitches in tucks or slipstitch structures – this should be taken into consideration.

Measurements of centre-back length, cuffs and sleeve length are taken either from a person, from a garment that already exists or from a jersey toile. Now work out how many stitches need to be knitted to create the required pattern pieces for the garment shape. Trims such as collars, cuffs and waistbands will all have to be knitted also.

Once fashioned pieces have been knitted they may need to be blocked before constructing together, as they may have lost their shape while on the machine. Chunky fabrics and ribs do not need to be blocked. Blocking involves placing the knitted piece face down on to a padded stiff backing, straightening the knit and checking the measurement. The knitted piece then requires careful steaming. Do not put the iron in direct contact with the knit as this may flatten stitches. Let the piece cool down completely before removing it from the backing. Fully fashioned knitwear can be sewn together by hand using a variety of hand stitches including back, blanket, overlock, or zigzag stitch on a sewing machine, which allows for more stretch than a normal straight stitch. Make sure you use a roller foot or cover the foot with tape so it doesn't get caught in the knit stitches if using a sewing machine. There should be as much stretch at the seams as there is in the rest of the garment. Pin together first so pieces do not overstretch, you can use the same yarn that the garment is made in so the stitches do not show. Use a blunt thick needle when hand sewing so that the yarn in the knit stitches is not split.

1   Vintage Junya Watanabe wool jersey t-shirt with silk, cotton and polyester appliqué flowers.

2   Peter Jensen fully fashioned neck detail (A/W07).

3   Louise Goldin machine-knitted dress (A/W07).

4–5   Junya Watanabe two-layer cardigan. This piece clearly shows full fashioning around the shoulders and neck.

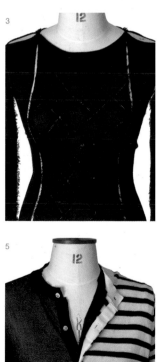

Woven fabrics

1

Transparent fabrics

Transparent fabrics allow seams, facings, hems and construction to show on the outside of a garment. Crisper transparent fabrics such as organza are easier to cut and sew than the softer transparents such as chiffon and georgette. Also the heavier the transparent is the easier it is to construct into a garment. Difficult fabrics easily distort when cutting and sewing. To help avoid this lay the fabric on the cutting table to rest overnight, roll it out flat on tissue paper with no overhang, then pin the fabric to the paper and carefully cut out the pattern pieces. The tissue can be left on the pieces while they are sewn together and then carefully ripped off after construction. The selvedge of the fabric can be successfully used as a finish of a seam. The seams used can affect the drape and structure

of the garment so try testing different types of seams first to see what works best. Very light transparent fabrics are often French seamed, as the seams can be seen from the outside this type of construction is very desirable.

Darts can be very hard to achieve successfully; maybe consider how seams can be placed to avoid darts. Leave garments to hang before hemming them to allow the fabric to drop, the hem can then be measured from the floor up, trimmed and sewn. Iron transparent fabrics at a lower temperature than heavier fabrics made from the same fibres.

French seams

To create a French seam, place the wrong sides of the fabric together and sew on the seam line. Next, trim the seam allowance down, press the seam open to flatten and then press the seam allowance to one side. Put the right sides together and sew a line of stitching that traps in the seam allowance.

Pile fabrics

Check the direction of piled fabrics, there might be an up and down that will create a slight tonal colour difference on pattern pieces and garments. When cutting fabric, mark the pile direction on the back of the fabric or on the selvedge edge and then lay all the pattern pieces in the same direction. Pile fabrics tend to shift when they are right side to right side, so lay fabric out wrong side together or completely flat. When pinning pattern pieces to velvet put the pins in the seam allowance as they can mark the fabric. To avoid fabrics shifting when you are creating seams hold fabrics taut when stitching or stitch with tissue in between the fabric. Be careful when pressing these fabrics as it can spoil the pile, so press from the wrong side only.

Outerwear fabrics

Seams used for outerwear fabrics often need to be strong and durable; a welt seam, which features topstitching is often chosen. Running a plastic coating along the seam from the inside can create a waterproof seam.

1   Comme des Garçons blouse. Cotton calico lining can be seen through the transparent viscose outer layer, enabling the internal and external elements of the garment to work together.

2   Jasper Chadprajong menswear designs.

2

Non-woven fabrics

Leather and suede

Leather and suede skins do not come as a length, but are specific to the size of the animal they come from. Think how big your pattern pieces are; you may have to create more seams in a garment if the skins are not big enough. Seams are easier to sew than darts so consider how you can position your seams. When choosing skins check them for imperfections, thinning and holes that you might have to work around, also compare skins for colour matching and pattern. Suede may have a nap so this must be considered when choosing skins and when sewing them up.

When cutting out use weights to hold down the pattern pieces, as pins will mark the leather. Wedge-point needles or leather needles cut cleanly through leather and suede, while a normal needle tends to rip the skin as it punctures the hard surface. Needle holes cannot be removed so take care sewing seams especially topstitching. It may be necessary to put a leather or Teflon foot on the machine to stop the leather sticking while sewing.

When sewing leather, press out seams from the wrong side and if they do not lie flat, glue the seam allowance down and lightly hammer from the back, being careful not to mark the leather on the right side. A very heavy leather might best be sewn with an overlapped seam. Remove seam allowance from one of the sides of the seam and overlap the other and topstitch down. It might be necessary to stick the surfaces first so they do not slip. If the leather is heavy, slash and glue darts or topstitch or overlap to create a flat finish. Leather stretches, but does not return to shape so it may be necessary to tape seams that are under strain from wear and tear.

Fur

Fur can be bought as tails, pelts or fur plates. Fur plates are fabrics created by sewing together smaller scraps of fur. When buying fur consider the size of your pattern pieces. Look for an overall good even coverage and colouration, pull the fur to see if it comes out, also see how soft and flexible the skin is as it may have dried out and not be worth buying. It might be possible to patch any holes and splits on the under leather side, as long as it doesn't affect the look on the fur side. It may also be possible to recycle dated second-hand fur garments into something new.

When cutting out fur lay right side down and place the pattern pieces in reverse on top, then cut through the leather skin and pull apart the fur or hair. Try not to trap long hairs or fur into seams, push them away from the seams before sewing. After seaming cut away fur or hair from the seam allowance to remove bulk and give a better finish.

Plastics

Plastics should be treated in a similar way to leather, as they may show needle holes, so take care when stitching. Be careful when pressing that you do not melt the fabric.

1   Jessie Lecomte A/W07.

How will you work?

As a textile designer you will need to consider the area you would like to specialise in, be it print, knit, weave, stitch or embellishment. You might work as a freelance designer producing samples that you sell yourself or through an agent. Alternatively you could work within a company designing textiles and also recolouring and preparing artwork for production. Textile ideas could also be taken into fashion garments, creating unique fabrics for fashion collections. There are also openings for textiles designers in the field of trend prediction, producing textiles for use within the trend prediction packages or working on trend ideas and colour. Here various textile designers have been interviewed to provide an overview of some of the jobs a fashion textile designer can have.

1

James Stone – Embroidery designer and part owner of Code Studio

What is your job title?
Embroidery designer.

Please describe your job.
I design one-off embroidered and beaded designs for the fashion industry. I personally design, produce and sell the designs to fashion companies worldwide.

Who have you worked for?
I have sold work to many companies around the world including Calvin Klein, Donna Karan and Nicole Farhi. I mainly design the work and show a portfolio, but sometimes I work on special commissions for runway development and concept.

What was your career path to your current job?
I studied at Norwich School of Art and then did an MA in Textiles at the Royal College of Art in London. I graduated in 1998 and then set up Code Studio with two friends. We now do two shows in Paris (at Première Vision) and three in New York a year.

What do you do on an average day?
I start work at 8.00am and normally spend most of my time designing for the next season. I also have appointments to arrange when needed and all the administration

work that goes along with working for yourself. I also have to source new fabric and yarn to sew with, which can mean shopping in fabric shops and antique markets.

What are your normal working hours?
I normally start at 8.00am and go through to 7.00pm each day. Sometimes when I have a deadline or a show coming up, I often work a little more.

What are the essential qualities needed for your job?
To be motivated, have enthusiasm, enjoy what you do and also strive to be creative with each design.

What kind of wage can someone command in your job?
There is an average price for a design throughout the textile industry. It is about £325 a design and the price can be more for larger pieces. You can also work on commission per day or sometimes the client gives you a budget and you explain what can be done in that time.

How creative a job do you have?
Very creative, I feel lucky to earn money from something I enjoy doing. You do have to work hard, but it's worth it in the end to be able to be creative all day.

What kind of team do you work with?
I did share a studio space with Emily and Lisa from Code Studio, but I now work from home, so I tend to design alone and then have lots of time with the clients whilst selling my work.

What is the best bit about your job?
It has been nice to travel with work and get a real insight into the industry from a freelance side. I also like to be able to push myself and work when and where I want.

And the worst?
The worst can be waiting to get paid from some companies, also having no financial support if things did not go well, but the good always outweighs the bad.

Any advice you would give someone wanting to get a job in your area of fashion?
Believe in yourself, always follow up any lead or interest there is in you and your work. Often you need to give something a go once or twice and then hopefully it could work out for a future design. Work hard at college, take full advantage of the facilities and workshops you have.

Duncan Cheetham – Printed textile designer

What is your job title?
Printed textile designer and consultant.

Please describe your job.
Print designer and direction for menswear and womenswear fashion. Colour and trend predictions. Print designer for Liberty and working on seasonal collections for a range of global fashion designers.

Who have you worked for?
Liberty of London, Pringle of Scotland, Marc Jacobs, DKNY, Donna Karan, Temperley, Calvin Klein, Diane von Furstenberg, Converse, WGSN and Penguin Books.

What was your career path to your current job?
MA in Printed Textiles at the Royal College of Art. BA (hons) Fashion and Textile design from UCE Birmingham.

What do you do on an average day?
Design, draw, research for design and colour. Concepts for client's trends and directions.

What are your normal working hours?
9.00am–5.00pm Monday to Friday and some weekend work when necessary.

What are the essential qualities needed for your job?
Organisation, good colour and design concepts and research skills, CAD skills in design and repeat, and knowledge of working to tight deadlines.

What kind of wage can someone command in your job?
£30–40k.

How creative a job do you have?
I have a very creative job involving drawing, colouring and initial concepts for clients.

What kind of team do you work with?
I work on my own doing freelance projects and with other design studios. At Liberty I work in a team of four designers and one archivist (working on a seasonal print collection that is wholesaled out to other designers).

What is the best bit about your job?
Seeing your designs through to colour and production, and seeing your designs in fashion products. Working with other design studios and a range of different projects, from flooring to book illustration.

And the worst?
At Liberty working two years in advance of the season. Not having time to do more freelance work!

Any advice you would give someone wanting to get a job in your area of fashion?
To gain as much experience in industry as possible, be original and different. And don't irritate anyone, it's a small industry! And work hard.

1   James Stone textile design.

2   Duncan Cheetham textile designs for Liberty.

Garment construction > How will you work?

Sid Bryan – Knitwear designer

What is your job title?
Knitwear designer.

Please describe your job.
Freelance designer. I go into companies and work on knit, from commercial selling collections through to one-off showpieces.

Who have you worked for?
Many designers at all levels of the market from high street and technical sportswear through to high-end designers worldwide.

What was your career path to your current job?
I wanted to study fine art/painting after my foundation course, but failed to get on to a painting degree. Thankfully I was directed towards textiles where I could utilise all my passion for colour and texture at Buckinghamshire on the BA Textile Design and Surface Decoration course. From there I studied Fashion Knitwear at the Royal College of Art and started working freelance with Bella Freud and Alexander McQueen. From then it snowballed and through word of mouth I met more and more designers.

What do you do on an average day?
There isn't really an average day. I travel a lot, so when I am back in the office my days are often taken up with administrative work, catching up with paperwork and liaising with suppliers and clients. I might be meeting with a designer to create a brief, brainstorming with my team, drawing, knitting, sampling and creating technical support documents for suppliers. I could be sourcing vintage stitchwork from markets or using a trampoline to create giant plastic doilies.

What are your normal working hours?
Endless! I live near my studio so I get up at 6.30am in the morning and work until as late as possible. I work most weekends.

What are the essential qualities needed for your job?
You have to have a certain sort of technical brain to be a knitter and patience is essential as you have to do things over and over again until it's right. Because I work with so many different designers you need to be able to adjust and work with every sort of personality.

What kind of wage can someone command in your job?
You can charge as much as you want, but ultimately you have to get a reputation as being reliable and talented and then decide how much you think that is worth.

How creative a job do you have?
Extremely, every time you create a piece of knit you start with a strand, a yarn then through an endless variety of processes you create a sample cloth and with that you then have an endless variety of shapes and forms this can take. There are joyous, really creative bits like knitting with elastic bands for the last Giles show and a lot of boring business and admin bits too.

What kind of team do you work with?
I have two full-time assistants who I couldn't function without and any number of people working with me as and when necessary. Last season we swelled to 12 at its peak.

What is the best bit about your job?
I knit, which is what I love and I get to work with amazing, talented people.

And the worst?
I travel so much that sometimes I don't know what day it is and I miss my husband and my dog.

Any advice you would give someone wanting to get a job in your area of fashion?
Ultimately you need to have a real passion and talent for what you do. There are no conventional routes, work with people and gain as much experience as you can. You must have a real technical understanding of knitwear and with this understanding push all the boundaries possible whenever possible. Above all, as with all of the creative industries be nice, be keen, it will go a long way.

Justine Fox – Colour and fabric trend prediction

What is your job title?
Project manager/colourist/fabric editor.

Please describe your job.
Working closely with clients to create colour palettes and forward trends for a range of industries including dyestuff, plastics, paints and electronics. I work in conjunction with graphic designers to come up with effective marketing solutions for B2B and B2C. I also oversee production from panel meeting to publishing of the Mix interior colour forecasting book. I identify, collate, style and write interior fabric pages in Mix Future Interiors magazine and I promote the company through presentations and workshops around the world.

Who have you worked for?
Global Color Research Limited. Clients include DuPont, Clariant, Homebase, KTF Retail, Comex and Addis.

What was your career path to your current job?
Studying Fashion BA at the University of Brighton, working in the fashion industry, interiors and branding, furniture and finally, colour psychology and colour trends.

What do you do on an average day?
A lot on client liaison, but it depends on the time of year – the industry is quite seasonal.

What are your normal working hours?
9.30am–6.00pm are contracted, but there is a lot of overtime particularly at the end of projects, when the trade fairs are on and business trips.

What are the essential qualities needed for your job?
A good understanding of colour, diplomacy, stamina, time management and at least basics in design packages.

How creative a job do you have?
I would say it's 40/60 between creative and administrative.

What kind of team do you work with?
A lot of freelancers – we choose the team dependent on the job.

What is the best bit about your job?
Travelling, fun clients and the finished product.

And the worst?
The hours and pay.

Any advice you would give someone wanting to get a job in your area of fashion?
Keep up to date on developing trends, go to as many exhibitions and trade shows as possible and be enthusiastic about colour.

Pastels   2

Naturals

1   Sid Bryan show piece.

2   Justine Fox colour palettes.

Garment construction > How will you work?

Antoine Peters – Fashion textile designer

What is your job title?
Initiator and creative director of the Antoine Peters design label.

Please describe your job.
Concept, design, development, technical and business direction.

Who have you worked for?
Work experience at Viktor & Rolf and *AvantGarde* magazine.

What was your career path to your current job?
Commercial economy, HEAO, Arnhem; Arnhem Institute for the Arts, Fashion Department; working at Viktor & Rolf and *AvantGarde* magazine; a master course from Fashion Institute Arnhem; self-employed fashion designer, design label Antoine Peters 'A sweater for the world!'; guest teacher ArtEZ, The Arnhem Academy of Art and Design, fashion design and product design.

What do you do on an average day?
Emailing, designing, pattern drawing and getting inspired.

What are your normal working hours?
9.00am to as late as needed.

What are the essential qualities needed for your job?
The same as those of a midfield footballer: creativity, all-roundedness, keeping an overview, positive coaching, team worker, winning mentality, self-critical and perseverance. Oh yes, it's nice to have a little business sense off the 'pitch' too. And you must be amorous with the ball.

How creative a job do you have?
One of the most creative ones in the world, because of the mixture of all the different disciplines, restrictions and lack of limits at the same time.

What is the importance of textiles in your work?
Silhouettes are very important to me and so are textiles. And I believe (silk) jersey and denim can be just as interesting in high fashion as the stereotypical richer fabrics. And print design is an even more important part of my collections, because it can add a lot to a story and is very communicative.

What kind of team do you work with?
Work experience students, friends and soulmates.

What is the best bit about your job?
The possibility to bring some lightness into the world. Force negativity out of the picture and connecting people by means of a smile.

And the worst?
Finding decent quality foreign production facilities and/or an investor. Someone?

Any advice you would give someone wanting to get a job in your area of fashion?
Listen, watch and ask as much as possible in the beginning. And then keep as close as possible to your new 'self'.

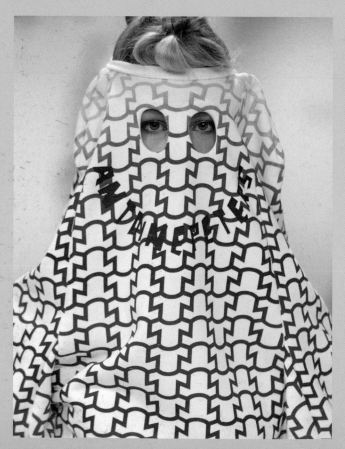

Manel Torres – Textiles with technology

What is your job title?
I am the managing director of Fabrican Ltd.

Please describe your job.
My role within the company is to give strategic direction and creative thinking within all areas the company operates (such as fashion, science and business) and ensure the financial buoyancy of the company.

What was your career path to your current job?
I studied for a PhD on the subject of spray-on fabric, which led me to found Fabrican Ltd.

What do you do on an average day?
Anything and everything.

What are your normal working hours?
As many as needed.

What are the essential qualities needed for your job?
Creativity, discipline and the ability to take risks.

What kind of wage can someone command in your job?
None of your business!

How creative a job do you have?
As stated before, creativity is essential in making the company what it is.

What kind of team do you work with?
I work with a team that has a varied background. It ranges from arts and design, to science and business.

What is the best bit about your job?
Creating products for future use, which no one has thought of yet.

And the worst?
Getting a fair agreement in a business deal.

Any advice you would give someone wanting to get a job in your area of fashion?
Persevere!

2

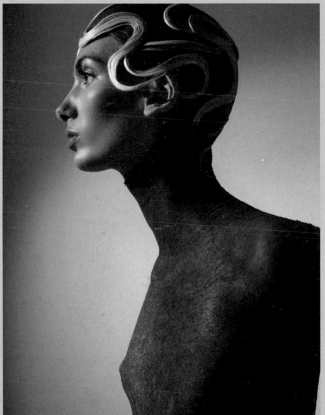

1   A SWEATER FOR THE WORLD! by Antoine Peters. Shown during Amsterdam International Fashion Week, July 2007.

2   Manel Torres textile design.

Garment construction > How will you work?

1   Vintage menswear. Dress
     shirt and waistcoat.

The intention of this book has been to try and cover all the areas relating to the research, design and creation of fashion textiles. The aim has been to give an insight into the topics a textile designer should know about in order to really understand their subject and to provide information fashion designers will benefit from to improve their textile work. This includes fibre qualities and fabric finishes, together with practical information on how to work, cut and sew textiles into garments.

I hope that the information about careers for textile designers in the fashion industry is useful. Quite often textile designers do not get credited for their amazing work, leaving the fashion designer to pick up all the acclaim. Textile design is often the unsung hero of fashion design, however, without innovative textiles fashion design would surely not be as interesting.

Fashion can be a difficult industry to work in, but also a very interesting and exciting one, so try to be the best you can. Use the information in this book to inform and stimulate your designs. Continue to research topics that have not been covered in detail here. It is important to really push your ideas, be innovative, experimental and above all take pleasure in what you are doing.

This book has been a challenge to compile, as there was so much to look at, the more I researched the more I wanted to know. I personally have learnt so much more about textiles through writing this book. I hope you get as much out of it as I have done.

Glossary

Brand image
Tangible and intangible characteristics that identify a brand.

Camouflage
Fabric originally developed by the French army during the First World War to disguise soldiers during field combat. The abstract coloured patterns have now been developed and adopted in fashion.

Classic
A garment that has a widespread acceptance over a period of time and is well known by name.

Colourfastness
How a fabric's colour reacts to washing, abrasion or light.

Colourways
Colour groups and combinations offered.

Computer-aided design (CAD)
The use of computers to design.

Drape
The way a fabric hangs.

Empire line
Dresses worn during the First Empire in France (1804–1815), which were characterised by a high waistline.

Engineered designs
Designs that are made to fit into a shape or are placed in a certain way.

Fashion week
Periods of time usually twice a year during which fashion collections are shown in the major fashion capitals of the world, for example, New York, London, Paris and Milan, to press and buyers.

Felted
The knotting together of fibres through heating with chemicals or friction to produce a matted material.

Finishes
Processes and techniques that are used to manipulate the appearance, characteristics, performance or handle of a fabric. Also used to describe the way a garment is neatened during construction, for example, with seams, hems and facings.

Fully fashioned
The shaping of a knitwear garment so each edge is a selvedge and will not unravel.

Handle
The tactile quality of a fabric.

Interfacing
Fabric placed between the garment and facing to add body, strength or structure.

Japonisme
The influence of Japanese arts on Western art.

Lining
Fabric used on the inside of a garment to hide the construction. It extends the garment's life as it helps to retain the shape and also makes the garment more comfortable to wear.

Mainstream
Trends that are accepted by the majority.

Mood boards
A collection of images, colours, objects or fabrics that are grouped together to express visually a theme or design idea.

Pattern pieces
The shapes that make up a garment in paper form, created through pattern cutting.

Penelope canvas
A double mesh canvas formed with pairs of crosswise and lengthwise intersecting threads.

Performance fabrics
High-tech fabrics that were originally developed for sportswear or extreme climate outerwear, but are now used for mainstream fashion.

Overlocking
Mostly used in knitwear for cut-and-sew production to cover cut edges and seams at the same time.

Smart textiles
Fabrics that respond to changes in their environment and alter in some way. Smart fabrics appear to 'think'.

Stand
A dressmaking mannequin or dummy.

Topstitch
To stitch on the right side of the garment.

Allen J (1989)
**John Allen's Treasury of
Machine Knitting Stitches**
David & Charles

Allen J (1985)
**The Machine Knitting Book:
How to Design and Create
Beautiful Garments on Your
Knitting Machine**
Dorling-Kindersley

Bendavid-Val L (ed.) (2004)
**In Focus: National Geographic
Greatest Portraits**
National Geographic Society

Black S (2002)
Knitwear in Fashion
Thames & Hudson

Braddock Clarke SE and
O'Mahony M (2005)
**Techno Textiles: Revolutionary
Fabrics for Fashion Design**
Thames & Hudson

Brannon EL (2005)
**Fashion Forecasting
(Second Edition)**
Fairchild Publications

Brittain J (1989)
**Needlecraft: Pocket
Encyclopedia**
Dorling-Kindersley

Campbell-Harding V and
Watts P (1993/2003)
Bead Embroidery
Batsford

Compton R (1983)
**The Complete Book of
Traditional Knitting**
Batsford

Cumming R and Porter T (1990)
The Colour Eye
BBC Books

Fogg M (2006)
**Print in Fashion: Design and
Development in Textile Fashion**
Batsford

Franklin T A and Jarvis N (2005)
Contemporary Whitework
Batsford

Fukai A (2002)
**Fashion: A History from the
18th to the 20th Century.
The Collection of The Tokyo
Costume Institute**
Taschen

Fukai A (2005)
**Fashion in Colors: Cooper-Hewitt
National Design Museum**
Assouline

Gale C and Kaur J (2004)
**Fashion and Textiles:
An Overview**
Berg

Gale C and Kaur J (2002/2006)
The Textiles Book
Berg

Gillow J and Sentence B
(1999/2006)
**World Textiles: A Visual Guide to
Traditional Techniques**
Thames & Hudson

Ginsburg M (1991)
The Illustrated History of Textiles
Studio Editions

Griffiths A (1989)
An Introduction to Embroidery
Mallard Press

Hencken Elsasser V (2005)
Textiles: Concepts and Principles (Second Edition)
Fairchild Publications

Holbourne D (1979)
The Book Of Machine Knitting
Batsford

Jackson L (2001)
Robin & Lucienne Day: Pioneers of Contemporary Design
Mitchell Beazley

Joyce C (1993)
Textiles Design: The Complete Guide to Printed Textiles For Apparel and Home Furnishing
Watson-Guptill

Kendall T (2001)
The Fabric & Yarn Dyer's Handbook: Over 100 Inspirational Recipes to Dye and Pattern Fabric
Collins & Brown

Kraatz A (1989)
Lace: History and Fashion
Thames & Hudson

Krevitsky N (1966)
Stitchery: Art and Craft
Reinhold Publishing

Lee S (2005)
Fashioning the Future: Tomorrow's Wardrobe
Thames & Hudson

McNamara A and Snelling P (1995)
Design and Practice for Printed Textiles
Oxford University Press

Meller S and Elffers J (1991/2002)
Textile Design: 200 Years of Pattern for Printed Fabrics Arranged by Motif, Colour, Period and Design
Thames & Hudson

Phaidon Press (ed.) (1998/2001)
The Fashion Book
Phaidon Press

Schaffer C (1994/1998)
Fabric Sewing Guide (Updated Edition)
Krause Publications

Seiler-Baldinger A (1994)
Textiles: A Classification of Techniques
Crawford House Press

Sorger R and Udale J (2006)
The Fundamentals of Fashion Design
AVA Publishing

Tymorek S (ed.) (2001)
Clotheslines: A Collection of Poetry and Art
Abrams

Watt J (2003)
Ossie Clark: 1965–74
V&A Publications

Wells K (1997/1998)
Fabric Dyeing and Printing
Conran Octopus

Bibliography

Useful resources

UK

British Fashion Council (BFC)
5 Portland Place
London
W1B 1PW

Tel: +44 (0)20 7636 7788
www.londonfashionweek.co.uk

*Owns and organises London
Fashion Week and the British
Fashion Awards.*

It also helps British fashion designers.

**The Department of Trade and
Industry (DTI)**
1 Victoria Street
London
SW1H 0ET

www.dti.gov.uk

*Business advice on export and legal
issues.*

Portobello Business Centre
www.pbc.co.uk

*Runs excellent business courses
for fashion designers and provides
advice on finding finance.*

Prince's Trust
www.princes-trust.org.uk

*Advice and finance for small
business start-ups.*

Fashion employment agencies
www.denza.co.uk
www.smithandpye.com

USA

**Council of Fashion Designers of
America**
www.cfda.com

*Promotes and advises American
fashion designers.*

Olympus Fashion Week
www.olympusfashionweek.com

*America's Fashion Week held in
New York.*

ITALY

**Camera Nazionale della Moda
Italiana (National Chamber for
Italian Fashion)**
www.cameramoda.it

*An association that disciplines,
co-ordinates and promotes the
development of Italian fashion. It
organises the Italian Fashion Week.*

FRANCE

**Fédération Française de la
Couture, du Prêt-à-Porter des
Couturiers et des Créateurs
de Mode**
www.modeaparis.com

*Under various departments it
oversees French couture and the
ready-to-wear design industry for
menswear and womenswear.*

*The federation organises the Paris
fashion shows and events at
Mode à Paris.*

JAPAN

**Council of Fashion Designers,
Tokyo**
www.cfd.or.jp

Promotes Japanese fashion.

HONG KONG

**Hong Kong Trade Development
Council**
www.hkfashionweek.fw.com

*Hong Kong Fashion Week is
organised by the Trade
Development Council.*
www.tdctrade.com

Useful resources

MUSEUMS

**Victoria and Albert Museum
(V&A)**
Cromwell Road
South Kensington
London
SW7 2RL
UK

www.vam.ac.uk

Fashion Museum
Assembly Rooms
Bennett Street
Bath, BA1 2QH
UK

www.fashionmuseum.co.uk

Costume Gallery
Los Angeles County Museum of Art
5905 Wilshire Boulevard
Los Angeles
CA 90036
USA

www.lacma.org

The Bata Shoe Museum
327 Bloor Street West
Toronto
Ontario
Canada
M5S 1W7

www.batashoemuseum.ca

**Museum at the Fashion Institute
of Technology**
7th Avenue at 27th street
New York
NY 10001–5992
USA

www.fitnyc.edu/museum

**The Costume Institute
The Metropolitan Museum of Art**
1000 5th Avenue at 82nd street
New York
NY 10028–0198
USA

www.metmuseum.org

**Museum of Fine Arts, Boston
Avenue of the arts**
465 Huntington Avenue
Boston
Massachusetts 02115–5523
USA

www.mfa.org

The Kyoto Costume Institute
103, Shichi-jo
Goshonouchi Minamimachi
Kyoto 600–8864
Japan

www.kci.or.jp

Textiles and Fashion

Kobe Fashion Museum
Rokko Island
Kobe
Japan

www.fashionmuseum.or.jp

MoMu
Antwerp Fashion ModeMuseum
Nationalestraat 28
2000 Antwerpen
Belgium

www.momu.be

Musée des Arts décoratifs
Musée des Arts de la mode et du
textile
107 rue de rivoli
75001 Paris
France
www.ucad.fr

**Musée de la Mode et
du Costume**
10 avenue Pierre 1er de serbie
75116 Paris
France

**Musée des Tissus et des Arts
décoratifs de Lyon**
34 rue de la charité
F-69002 Lyon
France

www.musee-des-tissus.com

Galeria del Costume
Amici di palazzo pitti
Piazza Pitti 1
50125 Firenze
Italy

www.polomuseale.firenze.it

Museum Salvatore Ferragamo
Palazzo Spini Feroni
Via Tornabuoni 2
Florence 50123
Italy

www.salvatoreferragamo.it

Wien Museum
Fashion collection with public library
(view by appointment)
A-1120 Vienna
Hetzendorfer
Straße 79

www.wienmuseum.at

Useful resources

Useful resources

FABRICS AND TRIMS

UK

Cloth House
47 Berwick Street, Soho,
London, W1F 8SJ
Tel: 0207 437 5155
www.clothhouse.com

Broadwick Silks
9–11 Broadwick Street, Soho,
London, W1F 0DB
Tel: 0207 734 3320

VV Rouleaux
54 Sloane Square, Cliveden Place,
London, SW1W 8AX
Tel: 020 7730 3125
www.vvrouleaux.com

Kleins
5 Noel Street, London W1F 8GD
Tel: 0207 437 6162
www.kleins.co.uk

Whaleys (Bradford) Ltd
Harris Court, Great Horton,
Bradford, BD7 4EQ
Tel: 01274 576718
www.whaleys-bradford.ltd.co.uk

US

NY Elegant Fabrics, NYC
222 West 40th St, New York,
NY 10018
Tel: +11 212 302 4980
www.nyelegantfabrics.com

M&J Trimming
1008 6th Ave, New York,
NY, 10018
www.mjtrim.com

WEBSITES

www.costumes.org

www.fashionoffice.org

www.promostyl.com

www.fashion.about.com

www.style.com

www.fashion-era.com

www.wgsn-edu.com

www.londonfashionweek.co.uk

www.premierevision.fr

www.hintmag.com

www.infomat.com

www.catwalking.com

Textiles and Fashion

PUBLICATIONS AND MAGAZINES

10	Marmalade
Another Magazine	Numéro
Arena Homme +	Oyster
Bloom	Pop
Collezioni	Selvedge
Dazed & Confused	Tank
Drapers Record	Textile View
Elle	View on Colour
Elle Decoration	Viewpoint
i-D	Visionaire
In Style	Vogue
International Textiles	W
Marie Claire	WWD Women's Wear Daily

FASHION FORECASTING

www.londonapparel.com

www.itbd.co.uk

www.modeinfo.com

www.wgsn-edu.com

www.peclersparis.com

www.edelkoort.com

FASHION TRADE SHOWS

www.premierevision.fr

www.informat.com

www.pittimmagine.com

www.purewomenswear.co.uk

www.magiconline.com

Useful resources

Useful resources

Alexander McQueen
www.alexandermcqueen.com

Balenciaga
www.balenciaga.com

Buddhist Punk
www.buddhistpunk.co.uk

Cathy Pill
www.cathypill.com

Chloé
www.chloe.com

Christian Wijnants
www.christianwijnants.be

Clare Tough
www.claretough.co.uk

Comme des Garçons
www.doverstreetmarket.com

Dior
www.dior.com

Dolce & Gabbana
www.dolcegabbana.it

Dries Van Noten
www.driesvannoten.be

Eley Kishimoto
www.eleykishimoto.com

Emma Cook
www.emmacook.co.uk

Givenchy
www.givenchy.com

Global Color Research Ltd.
www.globalcolor.co.uk

Hussein Chalayan
www.husseinchalayan.com

Jean-Pierre Braganza
www.jeanpierrebraganza.com

Jenny Udale
jennyudale@hotmail.com

Lecomte
www.jessielecomte.com

Jonathan Saunders
www.jonathan-saunders.com

Julia Krupp
www.julia-krupp.de

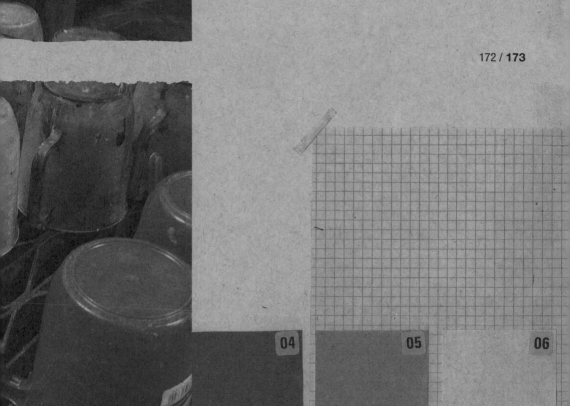

04 05 06

Kenzo
www.kozen.com

La Petite S*****
www.lapetitesalope.com

Louis Vuitton
www.louisvuitton.com

Manel Torres
www.fabricanltd.com

Manish Arora
www.manisharora.ws

Marc by Marc Jacobs
www.marcjacobs.com

Marios Schwab
www.mariosschwab.com

Marloes ten Bhömer
www.marloestenbhomer.com

Natalie Alabama Chanin
www.alabamachanin.com

Pantone
www.pantone.co.uk

Peter Jensen
www.peterjensen.co.uk

Peter Pilotto
www.peterpilotto.com

Prada
www.prada.com

Richard Sorger
www.richardsorger.com

Sandra Backlund
www.sandrabacklund.com

Sid Bryan
sid@sidneybryan.co.uk

Sonia Rykiel
www.soniarykiel.com

Tata-Naka
www.tatanaka.com

Viktor & Rolf
www.viktor-rolf.com

Wildlifeworks
www.wildlifeworks.co.uk

Woven
info@wovenstudio.co.uk

Useful resources

Canon

Sonia Rykiel

Manish Arora

Viktor & Rolf

Dior

Alexander McQueen

Prada

Hussein Chalayan

Givenchy

Dries Van Noten

Raf Simons

Siv Stoldal

Dior Homme

Martin Margiela

Balenciaga

Sandra Backlund

Marios Schwab

Christopher Kane

Marc Jacobs

Eley Kishimoto

Manel Torres

Jil Sander

Vivienne Westwood

Chloé

Issey Miyake

Junya Watanabe

Sophia Kokosalaki

Stella McCartney

Poiret

Coco Chanel

Dolce & Gabbana

Elsa Schiaparelli

Timorous Beasties

Mariano Fortuny

Paco Rabanne

Lucienne Day

Peter Jensen

Comme des Garçons

Emma Cook

Kenzo

Peter Pilotto

Christian Wijnants

Cathy Pill

Acknowledgements and picture credits

I would like to warmly thank the following people (in no particular order) for their contributions to this book:

Wildlifeworks, Marios Schwab, Peter Jensen, Clare Tough, Emma Cook, Daniela Bomba, Richard Sorger, Eley Kishimoto, Peter Pilotto, Julia Krupp, Christian Wijnants, Cathy Pill, Jessie Lecomte, Laura Miles at Woven, Antoine Peters, Sid Bryan, Le Petite S*****, Tata-Naka, Marloes ten Bhömer, Sandra Backlund, Manel Torres, Global Color Research Ltd, Duncan Cheetham, Timorous Beasties, Alabama Chanin, Spijkers en Spijkers, Jan Taminiau, Rory Crichton, Hannah Maughan, Michele Manz, Bruno Michel at At Large PR, Ibeyo PR, Village Press PR, Cube PR, A.I. press office and Sam Lewis.

I would like to especially thank Val Furphy for allowing me access to her amazing Comme des Garçons collection and her lovely textile samples. I thank Justine Fox for her intuitive colour work in Chapter five. James Stone for access to his collection of historical textile samples and his amazing embroideries, you are a very clever designer. Daniela Bomba for her striking cover image, many thanks. Thank you to Hywel Davies for your generous and kind support.

I would like to also thank Carl Downer and Daniel Stubbs at Sifer Design for their great work designing this book. All at AVA Publishing: Sanaz Nazemi, Brian Morris and Natalia Price-Cabrera, thank you for your help and support.

Finally much love to all my friends and family for always being there. Especially to my stylish Gran who is now in her 101st year. To my wonderful parents and to my long-suffering husband who has put up with my endless weekends working on this book, sorry.

Picture credits

Cover image courtesy of Daniela Bomba. Photography by Lindsay May; p 3 courtesy of Louise Henriksen; p 6 courtesy of Peter Pilotto and Sandra Backlund. Photograph by Peter Farago; p 10 courtesy of Catwalking.com; p 13 courtesy of Timorous Beasties; p 14–15 courtesy of James Stone; p 16–19 courtesy of V&A Images/Victoria & Albert Museum; pp 20–21 courtesy of James Stone; p 22 photography by Hywel Davies; p 27 courtesy of Laura-Maria Arola; p 28 courtesy of Jenny Udale; p 30 courtesy of Catwalking.com; p 31 courtesy of Jenny Udale; p 32 courtesy of Wildlifeworks; p 34 courtesy of Lousie Henrikensen; p 36 courtesy of Peter Pilotto and Sandra Backlund. Photograph by Peter Farago; p 37 courtesy of Alabama Chanin. Photograph by Peter Stangelmeir; p 38 courtesy of Wildlifeworks. Photographed by Natasha Cheek; p 39 courtesy of Catwalking.com; p 43 courtesy of Wildlifeworks. Photographed by James Parsons; p 44 courtesy of Christian Wijnants. Photography by Bettina Komenda; p 45 courtesy of Catwalking.com; p 47 courtesy of Sophie Copage; p 49 courtesy of Rory Crichton; p 52 photography by Lindsey May; p 53 courtesy of Marloes ten Bhömer; p 56 courtesy of Sam Lewis; p 57 courtesy of Catwalking.com; pp 58–59 courtesy of Catwalking.com; p 60.courtesy of Mika Nash p 63 courtesy of Furphy Simpson; p 64 courtesy of Catwalking.com; p 68 courtesy of Catwalking.com; p 73 courtesy of Spijkers en Spijkers. Photography by Marcio Madeira; p 74 courtesy of James Stone; p 77 photography by Lindsey May; p 78 courtesy of Catwalking.com; p 82 courtesy of Jenny Udale; p 83 courtesy of Christian Wijnants. Photograph by Bettina Komenda; p 86 courtesy of Marloes ten Bhömer; p 87 courtesy of Manel Torres. Photography by Miguel Dominguez; p 90 courtesy of Peter Pilotto. Photography by Pieter Huybrechts; p 93 courtesy of Cathy Pill. Photography by Etienne Tordoi; p95 images 2, 3 and 4 courtesy of Jenny Udale; p 96 courtesy of Catwalking.com; p 97 (2) courtesy of Furphy Simpson; p 99 (6–8) courtesy of Furphy Simpson; p 99 (9) courtesy of Louise Henriksen; p99 (10) courtesy of Spijkers en Spijkers. Photography by Chris Moore; p 100 courtesy of Catwalking.com; p 101 courtesy of Hannah Maughan p 102 courtesy of James Stone; p 103 (2) courtesy of James Stone; p 105 courtesy of Alison Willoughby; p 110 courtesy of Jessie Lecomte (photograph copyright jeanfrancoiscarly@jeanlouisjerome.com); pp 112–113 courtesy of Gavin Ambrose; pp 114–115 courtesy of Justine Fox. Copyright Global Color Research Ltd, all rights reserved; p 116 courtesy of Catwalking.com; p 117 courtesy of Catwalking.com, Copyright Global Color Research Ltd; p 118 (1) courtesy of Justine Fox. Copyright Global Color Research Ltd, all rights reserved; p 118 (2) courtesy of Catwalking.com; p 121 Courtesy and copyright of Global Color Research Ltd, all rights reserved; pp 124–125 courtesy of Sybarite; p 127 courtesy WGSN.com; p 128 courtesy of Catwalking.com; p 131 courtesy of Jessie Lecomte (photograph copyright jeanfrancoiscarly@jeanlouisjerome.com); p 133 courtesy of Jan Taminiau. Photography by Peter Stigter; p 134 courtesy of Catwalking.com; p 135 courtesy of Peter Pilotto and Sandra Backlund. Photograph by Peter Farago; p 136 courtesy of Catwalking.com; p138 courtesy of Catwalking.com; p 139 courtesy of Catwalking.com; p 140 (2) courtesy of Julia Krupp; p 147 courtesy of Cathy Pill and Jessie Lecomte (photograph copyright jeanfrancoiscarly@jeanlouisjerome.com); p 153 courtesy of Cathy Pill and Jessie Lecomte (photograph copyright jeanfrancoiscarly@jeanlouisjerome.com); p 154 courtesy of James Stone; p 155 courtesy of Duncan Cheetham/Liberty; p 156 courtesy of Sid Bryan; p 157 courtesy of Justine Fox. Copyright Global Color Research, all rights reserved; p 158 courtesy of Antoine Peters. Photography by Peter Stigter; p 159 courtesy of Manel Torres. Photography by Miguel Dominguez; pp 162–175 courtesy of Jenny Udale; pp 26, 42, 46, 50, 51, 61, 65, 66, 67, 70, 71, 72, 73, 75, 79, 80, 81, 84, 88, 91, 92, 94, 95, 97, 98, 101, 106, 108, 109, 119, 126, 137, 142, 143, 144, 145, 146, 148, 149, 150, 156, 158, 161, 166, 168, 172 photography courtesy of Andrew Perris Photography (www.andrewperris.co.uk).

Digger, Dumper and Dozer lived on a dirty patch of concrete.

The Boss made them
work very hard.

4

He didn't even let them
stop for a drink of water.

5

They had to build new
roads through the woods
and fields ...

... and knock down houses to make room for offices.

One day, the Boss said, "Tomorrow we're digging up that silly playground. We're going to build a factory instead."

9

"But where will the children play?" cried Dozer.

"Don't know, don't care," said the Boss.

Later, Digger said, "I'm not going to smash up the playground!"

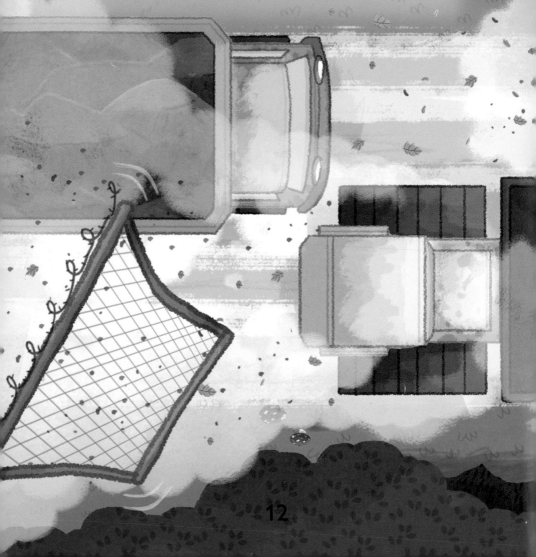

"Neither are we!" said the others. So they ran away to live in the forest.

"I know," said Dumper, "let's make a present for the children!"

It was very early in the morning as they sneaked into the town.

Everyone was still asleep.
So the machines built a
skateboard park ...

... and a duck pond.

They were making a
brand new playground
when suddenly ...

"Look! It's my machines! Stop them!" yelled the Boss.

"Run!" shouted Digger.
The machines drove
as fast as they could
towards the forest.

They only just got away!
The machines hid in
the trees.

But Digger was worried.
"We're bright yellow!
We're far too easy to see!"

Dozer said, "So let's paint ourselves green!" And they did!

They were just in time.
The Boss walked straight
past them and didn't
even notice.

Dumper whistled.
"That was close!"

26

Digger laughed. "While the Boss is looking for us, why don't we go back to the town and finish our new park?"

Later, Digger, Dumper and Dozer had gone back to the forest. But they left a note.

Have a great time, everybody! Love from The Green Machines XX

Puzzle 1

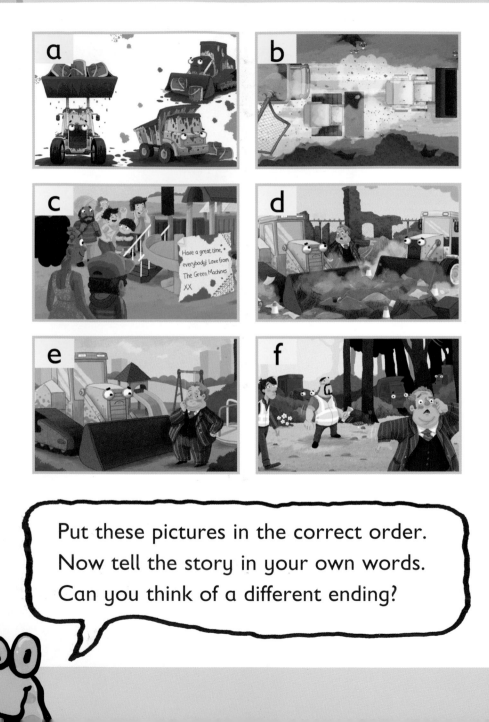

Put these pictures in the correct order.
Now tell the story in your own words.
Can you think of a different ending?

Puzzle 2

mean   selfish

generous

kind   greedy

considerate

Choose the words which best describe the characters. Can you think of any more? Pretend to be one of the characters!

Answers

Puzzle 1

The correct order is:

1d, 2e, 3b, 4a, 5f, 6c

Puzzle 2

Boss   The correct words are mean, selfish.

The incorrect word is generous.

Machines   The correct words are considerate, kind.

The incorrect word is greedy.

Look out for more stories:

For details of all our titles go to: www.franklinwatts.co.uk